MU120
Open Mathematics

Unit 11

Movement

MU120 course units were produced by the following team:

Gaynor Arrowsmith (Course Manager)
Mike Crampin (Author)
Margaret Crowe (Course Manager)
Fergus Daly (Academic Editor)
Judith Daniels (Reader)
Chris Dillon (Author)
Judy Ekins (Chair and Author)
John Fauvel (Academic Editor)
Barrie Galpin (Author and Academic Editor)
Alan Graham (Author and Academic Editor)
Linda Hodgkinson (Author)
Gillian Iossif (Author)
Joyce Johnson (Reader)
Eric Love (Academic Editor)
Kevin McConway (Author)
David Pimm (Author and Academic Editor)
Karen Rex (Author)

Other contributions to the text were made by a number of Open University staff and students and others acting as consultants, developmental testers, critical readers and writers of draft material. The course team are extremely grateful for their time and effort.

The course units were put into production by the following:

Course Materials Production Unit (Faculty of Mathematics and Computing)

Martin Brazier (Graphic Designer)
Hannah Brunt (Graphic Designer)
Alison Cadle (TEXOpS Manager)
Jenny Chalmers (Publishing Editor)
Sue Dobson (Graphic Artist)
Roger Lowry (Publishing Editor)
Diane Mole (Graphic Designer)
Kate Richenburg (Publishing Editor)
John A.Taylor (Graphic Artist)
Howie Twiner (Graphic Artist)
Nazlin Vohra (Graphic Designer)
Steve Rycroft (Publishing Editor)

This publication forms part of an Open University course. Details of this and other Open University courses can be obtained from the Student Registration and Enquiry Service, The Open University, PO Box 197, Milton Keynes MK7 6BJ, United Kingdom: tel. +44 (0)845 300 6090, email general-enquiries@open.ac.uk

Alternatively, you may visit the Open University website at http://www.open.ac.uk where you can learn more about the wide range of courses and packs offered at all levels by The Open University.

To purchase a selection of Open University course materials visit http://www.ouw.co.uk, or contact Open University Worldwide, Walton Hall, Milton Keynes MK7 6AA, United Kingdom, for a brochure: tel. +44 (0)1908 858793, fax +44 (0)1908 858787, email ouw-customer-services@open.ac.uk

The Open University, Walton Hall, Milton Keynes, MK7 6AA.

First published 1996. Second edition 2008.

Edited, designed and typeset by The Open University, using the Open University TEX System.

Printed and bound in the United Kingdom by The Charlesworth Group, Wakefield.

ISBN 978 0 7492 2869 9

2.1

Contents

Study guide

This unit builds on the work of *Unit 10*, introducing you to quadratic functions, another type of mathematical function for your modelling library.

Section 1 introduces some terminology, words such as *position*, *velocity* and *acceleration* used in a mathematical context. It also looks at position–time graphs, which are similar to distance–time graphs, and includes discussion of the phenomena of speeding up and slowing down, for which quadratic functions are often appropriate models.

You may have met some of these ideas before, not least if you have watched the television programme *Designer Rides*, which is very relevant to this unit, as it gives practical illustrations of many of the concepts of motion. You may find it useful to look at the notes for this programme after your study of Section 1.

Section 2 looks in more detail at quadratic functions, from an algebraic perspective, and uses the course calculator to fit a quadratic function to given data. This builds upon the work you did on linear regression in *Unit 10*, and there is extensive use of the *Calculator Book*. The concepts in this section are very important, as they generalize to other functions. You should make sure you study it thoroughly. The time it takes you will depend upon your algebraic and calculator skills.

In using a quadratic function, you will often need to find when the function takes a particular value, which can be done by solving a quadratic equation. Section 3 looks at such equations and at different ways of solving them, both graphically and algebraically, some of which involve using your calculator.

Section 4 is a video section. It builds upon the work of Sections 1 to 3, but should you need to view it earlier in your study, you will probably be able to understand most of it, except for how to handle the quadratic equation, after your study of Section 2.

Note that the material in the Appendix is *optional*.

As you work through this unit, you are asked to think about how you tackle difficult ideas—for example, if you get 'stuck' over a particular idea, how do you go about getting 'unstuck'. Record your thoughts on the Learning File sheet supplied.

Also, as you work through the unit, record the meanings of all new terms on the Handbook sheet provided. Be particularly careful to record the *mathematical* meaning of those terms whose mathematical use differs from their everyday one.

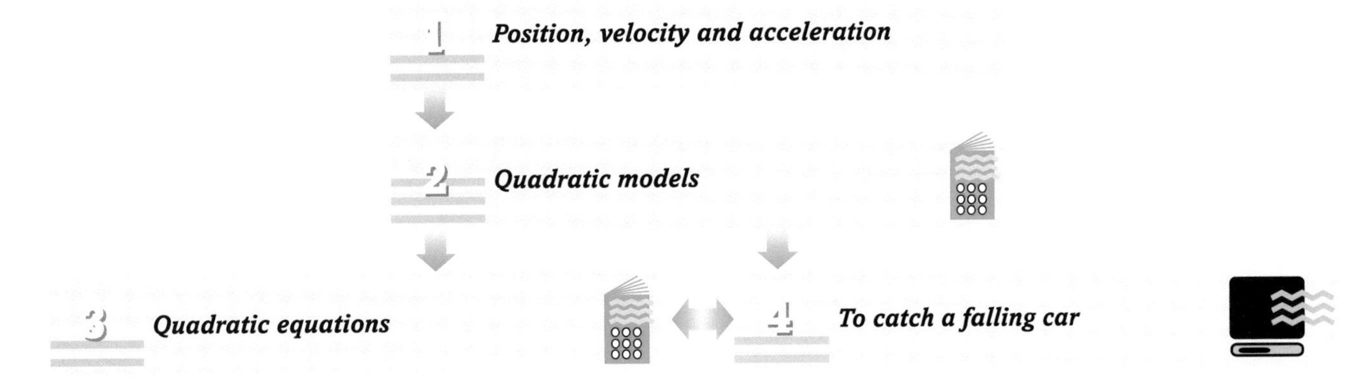

Summary of sections and other course components needed for *Unit 11*

Introduction

In *Unit 10*, you met models of motion based upon average speed, which produced straight-line distance–time graphs. While such models are appropriate for some situations, they do not provide a good fit where the purpose of the model includes an analysis of speeding up or slowing down.

Two concepts are particularly important in modelling situations that include an analysis of speeding up or slowing down. The first is *velocity*: velocity, rather than speed, is used in situations where the direction of the motion may change, for example a ball being thrown upwards and then coming down again. The second is *acceleration*: acceleration is used in a rather broader sense mathematically than in everyday life; in mathematics, it refers simply to a *changing* velocity—as well as indicating speeding up, it can also mean slowing down, depending on the direction of the acceleration. This is an example where, when speaking mathematically, you need to use some terms in a different way from in everyday speech.

The distinction between velocity and speed was mentioned in *Unit 7*.

Armed with these concepts, the unit will look at a different model for motion based upon average acceleration rather than average speed, which results in a straight-line velocity–time graph rather than a straight-line distance–time graph.

You have already met the x^2 key on your calculator: $y = x^2$ is the simplest quadratic function. The unit shows how this particular function is related to other functions all of whose highest power of x is a square term: the *quadratic functions*. It also shows how quadratic functions can be used to model a variety of situations, including ones involving an analysis of speeding up or slowing down.

As in previous units, the models in this unit stress certain aspects and ignore others. One situation modelled here is that of falling objects; this culminates in the video band 'To catch a falling car', which shows the use of quadratic functions in the design of the way a car airbag opens after a crash in order best to protect the occupants' heads.

1 Position, velocity and acceleration

Aims This section aims to introduce the mathematical ideas of velocity and acceleration, and to show how they are related to position–time graphs for objects which are speeding up or slowing down. ◇

You have by now become familiar with assuming a constant speed for models involving motion: the motion of a car driving along a motorway, the motion of a ferry crossing the English Channel, the motion of a train between stations. However, cars, ferries and trains do not reach their top speed immediately they start off on their journey. Starting from a stationary position (that is, moving with zero speed), they accelerate and their speed increases accordingly. In order to make predictions about such motion, it is helpful to be able to represent it clearly and unambiguously, and mathematical models provide a means of doing this. However, remember that mathematical models stress certain aspects of a situation while ignoring others.

This section introduces mathematical terminology in relation to everyday experiences of motion, and illustrates these ideas graphically; an algebraic representation is provided in the following section. Up until now, the direction of motion has often not been relevant, but in this section you will be looking at situations where it is.

1.1 Distance, position and direction

Imagine that you are standing upright and take one step forwards and then one backwards to your original position. You have moved a *distance* of two steps, but your final *position* is the same as before you started moving. This is because the steps you took were in opposite directions. So when describing movement in which there is the possibility of moving in different directions, the *direction* of motion is important.

Distance and position are different ideas, and so this section will distinguish between distance–time graphs (which you have met before) and *position–time* graphs. Also this unit will look at motion in *one dimension* only, for example, forwards and backwards, or upwards and downwards, ignoring any simultaneous sideways motion.

Activity 1 Bus trip

Imagine you go on a bus trip to visit a friend who lives 10 km away. You make the half-hour bus trip there, spend about an hour with your friend and then make the half-hour bus trip back. Describe your journey, making particular reference to the distance travelled and your position:

(a) in words;

(b) graphically.

What have you stressed in your description and what have you ignored?

In order to describe your position unambiguously in Activity 1, you needed to specify an *origin*—this might either have been where you started your journey or your friend's house.

You also needed to make a decision about the direction. Did you take your starting point to be above or below your friend's house on the y-axis (the position axis) of your graph? If you chose to mark your friend's house above your starting point, as in Figure 1, then your outward journey would be in the positive y-direction (your position increasing from 0 km to 10 km), and your return journey would be in the negative y-direction (your position decreasing from 10 km to 0 km). (The arrow on the y-axis always indicates the positive direction, by convention.)

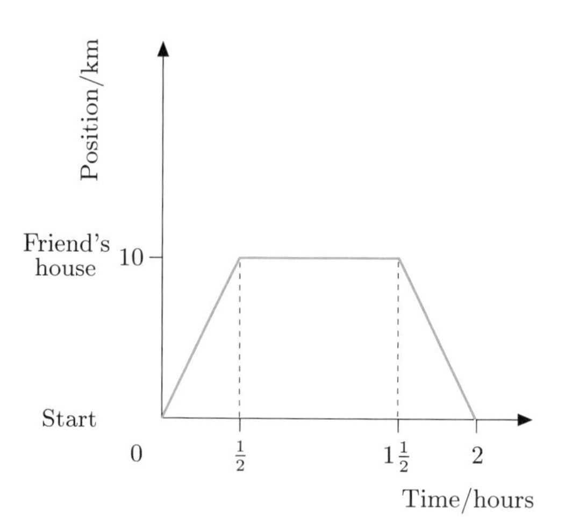

Figure 1

But if you chose to mark your starting point above your friend's house, as in Figure 2, then your position–time graph of the trip would have been different.

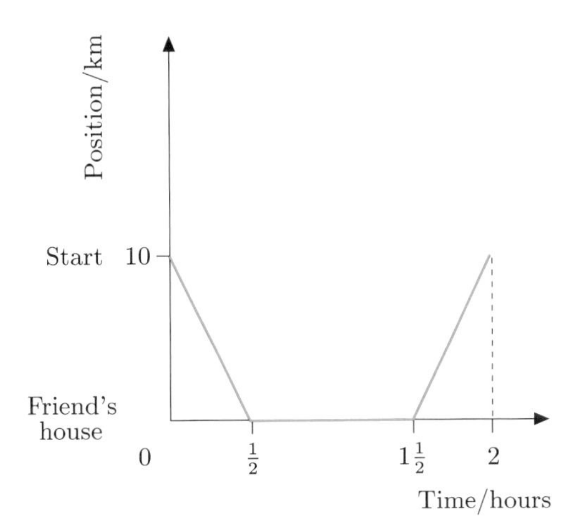

Figure 2

Both models are equally valid.

In general, when describing motion (in one dimension) mathematically, you must:

(a) choose the origin of the motion from which position is to be measured;

(b) choose which direction is to be considered the positive direction (not surprisingly, the opposite direction is then the negative direction).

Activity 2 Throwing a ball upwards

Imagine putting your arm out of a window and throwing a ball directly upwards. Draw a position–time graph to describe the ball's motion from the time it leaves your hand until it hits the ground.

In this case, it seems sensible to take upwards as the positive direction. However, you could choose either the throwing point or the ground as the origin for position. Both models are illustrated in Figure 3. The ball will first go upwards (in the positive direction), slow down, stop momentarily, and then fall downwards (in the negative direction) to the ground. The horizontal axis shows the time since the ball was thrown.

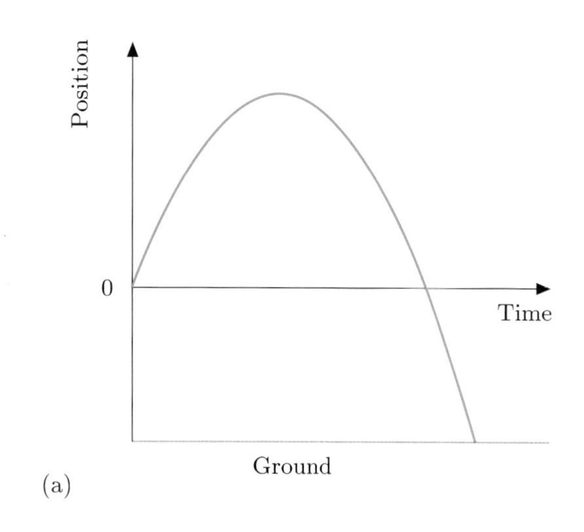

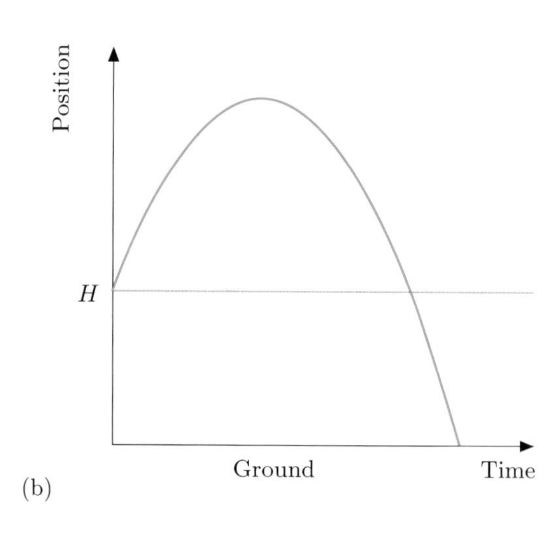

Figure 3

Although it could seem a bit contrary, the positive direction could have been taken as downwards, and so the graphs in Figure 4 are perfectly valid, if counter-intuitive, models!

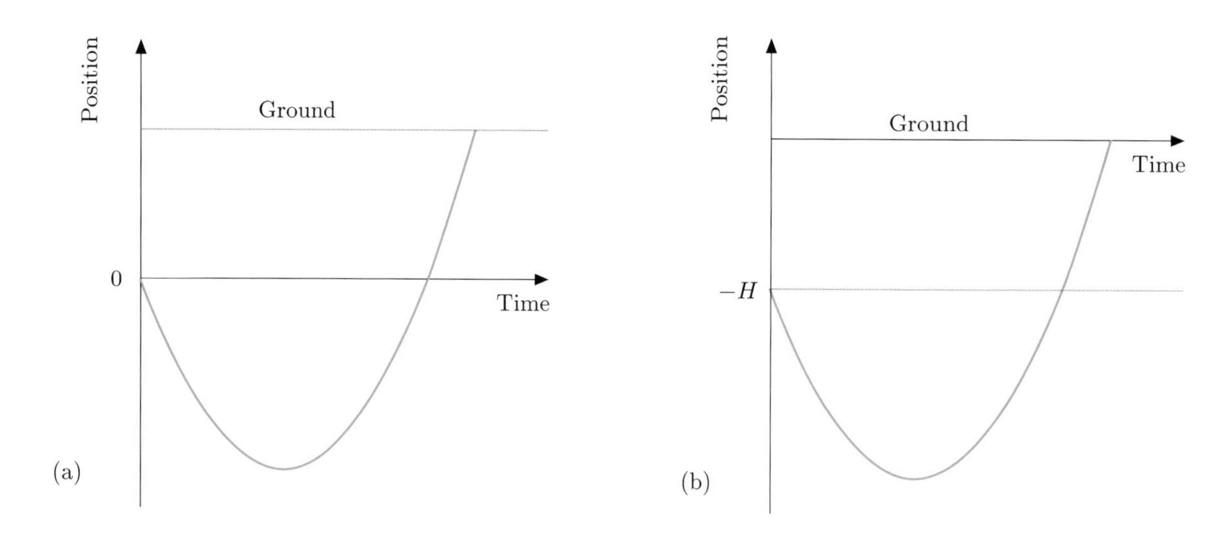

Figure 4

There is an important point here. There is a temptation to see position–time graphs as photographs or drawings of the actual situation—they are not. The images in Figure 4 look 'wrong' because of an erroneous sense of 'gravity' operating in such graphs. And even the sketches in Figure 3 do not show the ball's actual path: the ball goes straight up and down. Both figures are merely symbolic representations of reality.

1.2 Speed, velocity and direction

When you throw a ball straight up in the air, it leaves your hand with a certain speed, gradually slows down, momentarily stops, and then starts to fall back downwards, with increasing speed. Models that include the direction of motion as well as the numerical speed make use of the concept of *velocity*. Velocity is speed in a particular direction; and, when the motion is in one dimension, its value may be positive or negative to reflect the direction of travel. In any particular situation, one direction is specified as positive. Motion in that direction is indicated by a positive velocity. Since the opposite direction is specified as negative, motion in that direction is indicated by a negative velocity. Recall the bus trip in Activity 1. If the positive direction were taken to be *from* the starting point *to* your friend's house, then on the outward journey the velocity would be positive, but on the return journey it would be negative.

Mathematically, *velocity* is the rate of change of position, and is given by the slope or gradient of the position–time graph.

Figure 5 shows a ball thrown into the air from a person's hand, with the positive direction taken as upwards. Position is measured in terms of the height above the person's hand when the ball was thrown: call this height h. Positive values of h indicate heights above the hand and negative values of h heights below the hand.

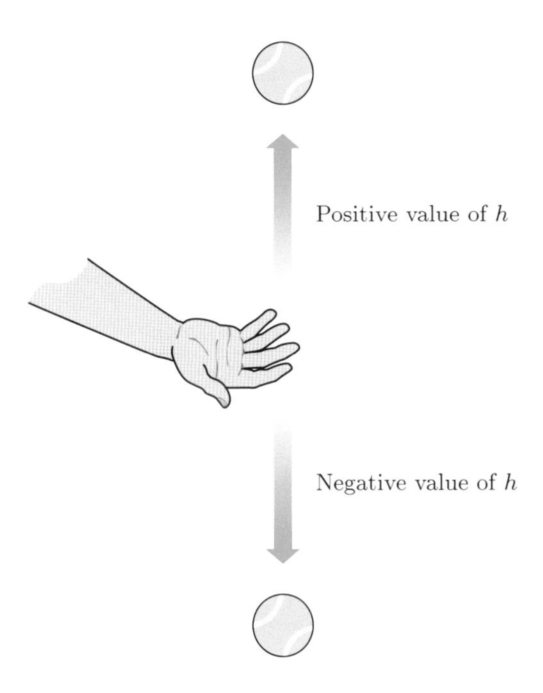

Figure 5

Positive values of the velocity v indicate speed in an upward direction and negative values of v indicate speed in a downward direction. Zero velocity indicates the ball is stationary, which happens at its highest point.

The position–time graph for the ball is obtained by plotting the height h against the time t, as shown in the top part of Figure 6. The slope or gradient of the graph gives the velocity. The slope is positive initially, zero at the highest point, and then negative. This slope is plotted against time in the bottom part of Figure 6, to give a *velocity–time graph* for the ball.

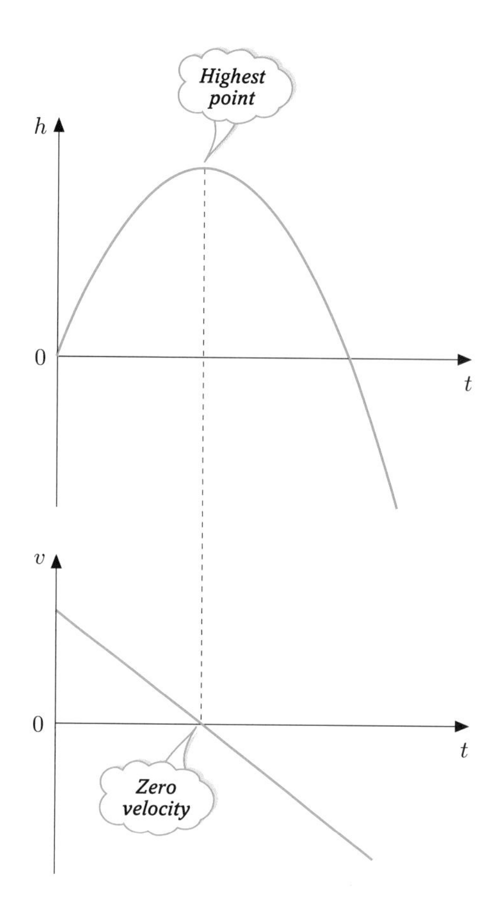

Figure 6

In *Unit 10*, the position–time graphs were all straight lines, as they were based upon average speed. These were constant-speed models, which ignored any speeding up or slowing down. If these phenomena are included in a model, then the position–time graph is not a straight line, but a curve. You know how to find the gradient or slope of a straight line, but what about the gradient or slope of a curve? First, consider *qualitative* descriptions and sketches before trying to produce specific numerical values and quantitative descriptions.

Figure 7 shows the position–time graph modelling the motion of a bus pulling away from a bus stop. Look at the curve.

► How would you describe its slope?

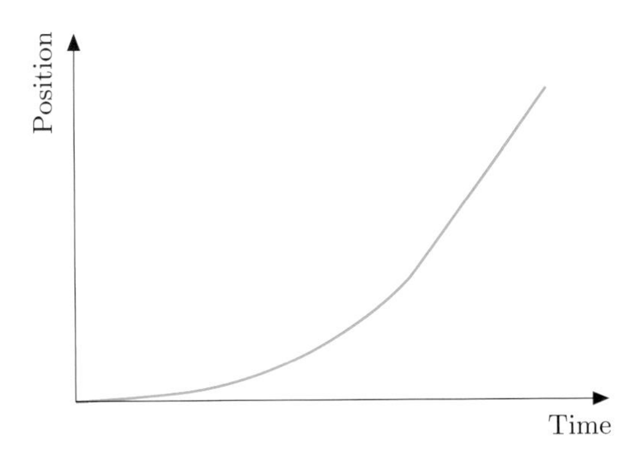

Figure 7

Initially, the slope is zero. Then it increases (reflecting the fact that the curve gets steeper). Then it becomes roughly constant. These changes correspond to zero velocity when the bus is stationary at the bus stop, increasing velocity, then constant velocity. The corresponding velocity–time graph is shown in Figure 8.

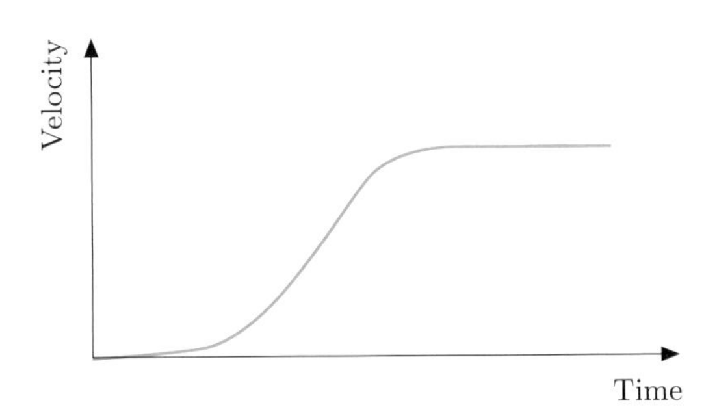

Figure 8

Activity 3 Slowing down

As a bus slows down, its velocity decreases. Sketch the position–time graph of a bus slowing down from a constant velocity to being stationary at a bus stop. By inspecting the gradient (slope) at different places on this graph, sketch the corresponding velocity–time graph.

The diagrams above and in the comment on Activity 3 do not show numerical values for the velocity (except for zero in some cases). They reflect a qualitative rather than a quantitative model. In many situations, however, quantitative models are required: for example, when timetabling trains, designing safety features of vehicles or planning the launch of a rocket into space.

To obtain a *quantitative* description of velocity from a position–time graph, you need numerical values for the gradient. To find the gradient at a particular point, draw a straight line through the point in the same direction as the curve. The line should just touch the curve at this point, as in Figure 9. Such a line is called the *tangent to the curve* at this point (from the Latin verb *tangere*, meaning 'to touch').

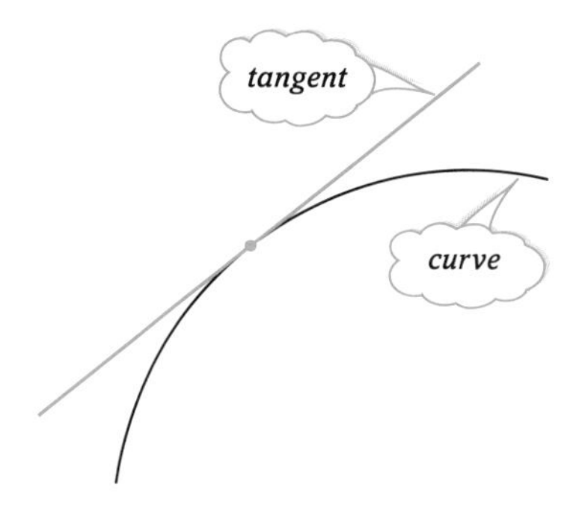

Figure 9 Tangent to a curve at a point

Example 1 Testing the brakes

In the design stage, many tests are carried out on vehicles and their braking systems. Such tests are often videoed, and from such videos important information can be deduced. Figure 10 shows a position–time graph of the front of a car as it brakes, which was obtained by using data from such videos. It can be used to find the velocity of the car during braking. Suitable tangents have been drawn to help you to visualize how the velocity is changing.

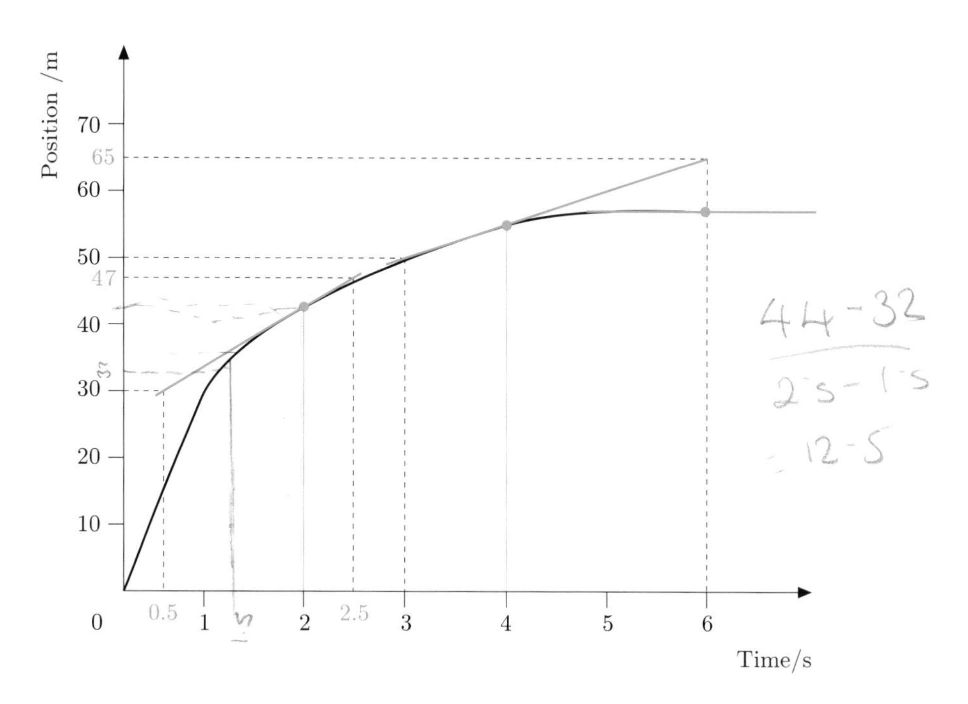

Figure 10

During the first second, the velocity is more or less constant. The gradient is 30 metres per second (sometimes written as $30\,\mathrm{ms}^{-1}$).

At 2 seconds, the gradient is obtained by drawing the tangent to the curve at the point where the time is 2 seconds. Since the tangent is a straight line, its gradient is obtained by choosing two arbitrary points on it, for example the points (0.5, 30) and (2.5, 47), and calculating:

$$\frac{\text{increase in position}}{\text{increase in time}} = \frac{47\ \mathrm{m} - 30\ \mathrm{m}}{2.5\ \mathrm{s} - 0.5\ \mathrm{s}} = \frac{17\ \mathrm{m}}{2\ \mathrm{s}} = \frac{17}{2}\ \mathrm{ms}^{-1} = 8.5\ \mathrm{ms}^{-1}$$

Similarly, at 4 seconds the gradient is:

$$\frac{65\ \mathrm{m} - 50\ \mathrm{m}}{6\ \mathrm{s} - 3\ \mathrm{s}} = \frac{15}{3}\ \mathrm{ms}^{-1} = 5\ \mathrm{ms}^{-1}$$

At 6 seconds, the gradient is zero and so the car has stopped.

This information can be represented on a velocity–time graph, as in Figure 11.

Remember there are two quite different uses for single letters: as algebraic variables and as abbreviations for units of measurement. Algebraic variables are normally italicized in order to mark this difference. However, notice that the notation s^{-1}, which means 'per second', treats the unit 'seconds' as if it were an algebraic quantity:

$$\frac{17\,\mathrm{m}}{2\ \mathrm{s}} = \frac{17}{2}\,\mathrm{m}\,\frac{1}{\mathrm{s}} = \frac{17}{2}\,\mathrm{ms}^{-1}$$

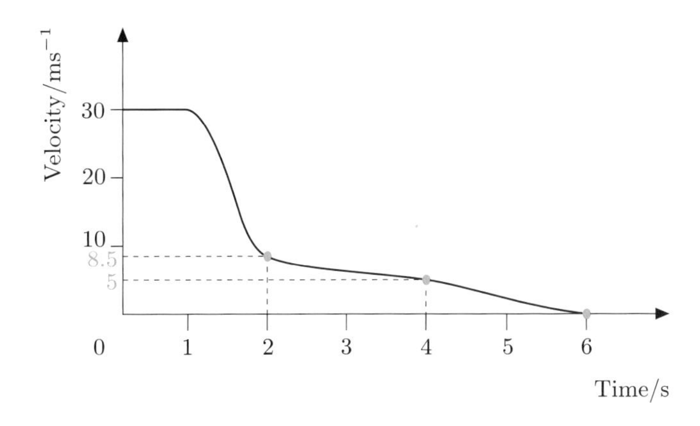

Figure 11

In order to obtain a more detailed velocity–time graph, tangents could be drawn at intermediate points, in particular between 1 and 2 seconds.

Activity 4 *Velocity after 1.5 seconds*

Draw in a tangent at 1.5 s, find its gradient, and hence work out the velocity at this time.

So velocity is the rate of change of position—and is given by the slope of the position–time graph. The value of the velocity tells how position is changing with time.

1.3 Acceleration, acceleration and direction

You may have thought that the repeated word in the subsection heading was a mistake. But it is not. It was put there to make a point, namely that there is no word in English to distinguish acceleration that takes direction into account and acceleration that does not.

If you look back to the previous two subsection headings, you will see that the first word in each ('distance' and 'speed', respectively) refers to a concept where direction does not feature. With these concepts, it makes no sense for them to have negative values, and a zero value means no distance or no speed.

The second word in each heading refers to a similar notion, but one where direction is taken into consideration, in these cases, 'position' and 'velocity'. Zero position does not mean that you are nowhere; rather, it means you are at a distinguished point from which position is measured, and so negative position makes sense. Zero velocity means you are not moving in either direction, and so again negative velocity makes sense.

Negative position and negative velocity are quite independent of one another. You can be at a negative-valued position and move in either the positive direction or the negative direction quite happily. Similarly, you can be at a positive-valued position and do likewise.

So now there is also acceleration. The basic real-world phenomenon of *acceleration* relates to something speeding up or slowing down. As was mentioned above, English does not have two different words to mark the distinction involving direction or not. When a vehicle speeds up, its speed is changing. It is said to be accelerating. However, the speed also changes when the vehicle slows down, as in Example 1 and Activity 3. In ordinary speech, the word 'decelerating' is sometimes used for slowing down, though the pedal in a car which causes it to slow down is called the brake, rather than the 'decelerator', which would parallel the adjacent pedal being called the 'accelerator'.

In mathematics, the word *acceleration* has a slightly broader meaning than in everyday life, where it simply means getting faster. In mathematics, acceleration means changing velocity, which occurs when something is getting faster *or* slower. Velocity usually changes over time. *Acceleration* is defined as the rate of change of velocity, and is directly reflected by the gradient of the velocity–time graph. A positive gradient is equivalent to positive acceleration, which means speeding up in the positive direction. A negative gradient is equivalent to negative acceleration, which means speeding up in the negative direction. So the phenomenon of 'slowing down' is seen mathematically as an acceleration in the opposite direction to the direction of travel, and allows the unfamiliar-sounding but mathematically correct statement 'the bus accelerated to a halt' to be meaningful.

This concept may take a little while to sink in. Try to remember that acceleration may also result in an object slowing down rather than speeding up! During the next few days, whenever you are moving about, try to think whether your velocity is changing, and if so consider in which direction you are accelerating.

For example, imagine that you are getting on a bus which sets off along a straight road. For convenience, take forwards as the positive direction. As the bus pulls away from the bus stop, it is accelerating with a forward (positive) acceleration. When it reaches its cruising speed, its velocity is constant, and it is not accelerating at all; it has zero acceleration. However, as it slows down for traffic lights or a pedestrian crossing, the velocity is changing again, and this time the bus can be described as having a negative acceleration. The position of the bus, however, is always positive. The velocity–time graph of a bus speeding up, cruising and then slowing down is shown in Figure 12.

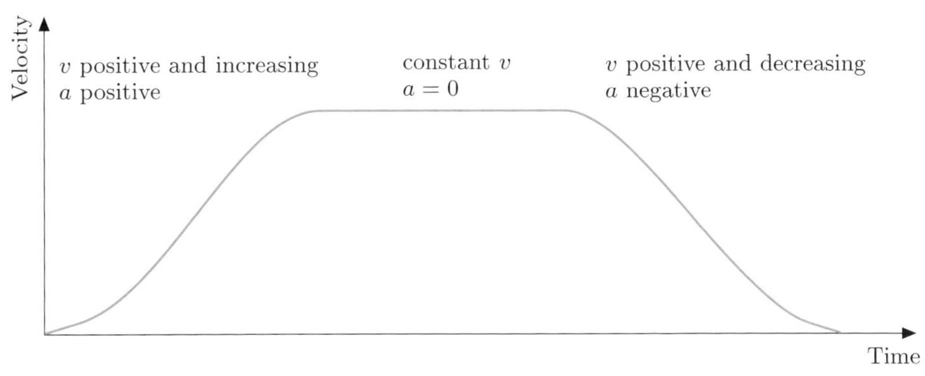

Figure 12

The velocity (v) is in the forward direction all the time, so the velocity is always positive. When the velocity increases, the acceleration (a) is positive; when it decreases, the acceleration is negative. Acceleration can be positive or negative quite independently of whether the velocity is positive or negative and of whether the position is positive or negative. (You might like to think through the mathematical description of a bus travelling in the opposite direction but carrying out the identical sequence of movements.)

As a passenger on the bus, you can tell that the bus is accelerating from the different directions in which you seem to be pulled relative to the bus. If the bus accelerates forwards (a positive value for the acceleration), you will seem to be left behind and feel a pull backwards. If the bus accelerates backwards (that is, it brakes, giving a negative value for the acceleration), you will seem to be being pushed forwards relative to the bus. If the driver keeps alternating rapidly between forward acceleration and backward acceleration, you get a rather uncomfortable ride. Smooth variations in acceleration make for a more comfortable ride.

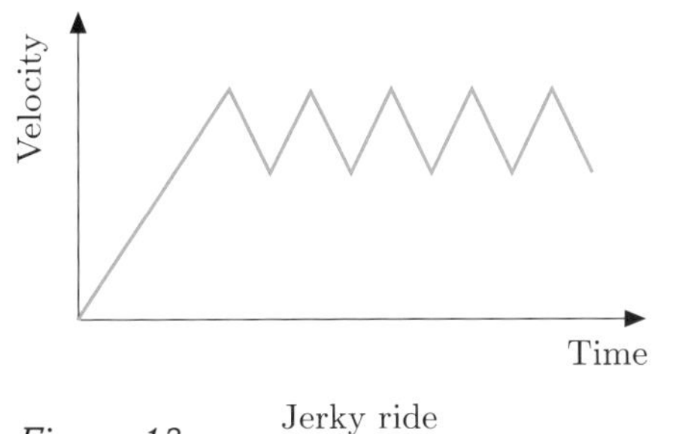

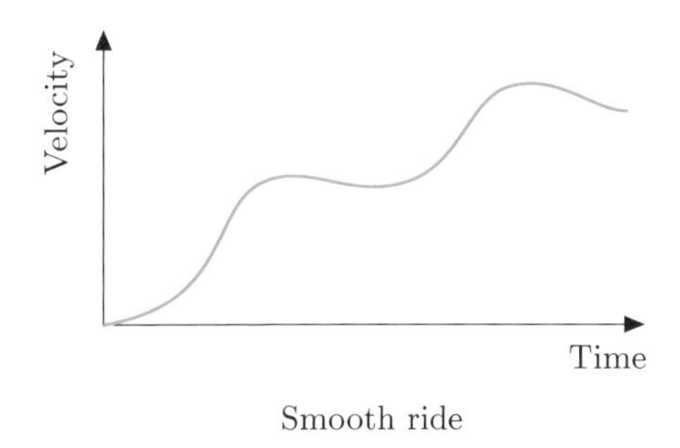

Figure 13

Activity 5 On the bus

Imagine you are still on the bus. Describe what happens to the acceleration and velocity as the bus pulls up at the next bus stop, picks up passengers and then starts off again. Sketch the velocity–time graph of the bus and indicate on the graph when the acceleration is positive and when it is negative.

This description assumes the rocket travels only in one dimension, in other words that it goes straight up and down.

You may have seen advertisements for cars telling you that the car can accelerate from 0 to 100 km per hour in so many seconds, to indicate its power of acceleration. A rocket firework accelerates upwards at its launch. It may have a period of constant velocity (zero acceleration) while it burns and displays wonderful colours, but then it slows down, thus accelerating in this opposite direction (negative acceleration), before it starts to fall to earth, with (negative) acceleration in this opposite direction.

Activity 6 Which are accelerating?

In which of the following situations is the object concerned accelerating?

(a) A car driving away from a traffic light.

(b) A person diving from a springboard into a swimming pool.

(c) A marathon runner running at a steady pace in the middle of the race.

(d) A train coming to a stop at the buffers.

(e) A spacecraft moving at a steady speed in a straight line far away from any gravitational influence.

Over short periods of time, it may be convenient to assume acceleration is constant. For instance, when the bus is speeding up you might assume it is steadily gaining speed. One example of where constant acceleration provides a good model is gravity. In the case of the ball thrown up into the air, Figure 6 showed a straight-line velocity–time graph, which implies a constant rate of change of velocity: that is, a constant acceleration. In the graphs in Figure 6, the positive direction is upwards; and, as the velocity–time graph gradient is negative, the acceleration is negative. This means that all the acceleration is in a downward direction.

▶ What does this mean?

As soon as the ball leaves the hand, it is being accelerated in the negative direction by gravity—that is, downwards. But its initial velocity is upwards—positive—and so its position changes initially in a positive direction. So the ball travels up into the air, but is continually slowing down, until it reaches its highest point; it then starts speeding up in a downward direction. Negative acceleration simply means acceleration in the negative direction, in this case downwards. This is illustrated in Figure 14.

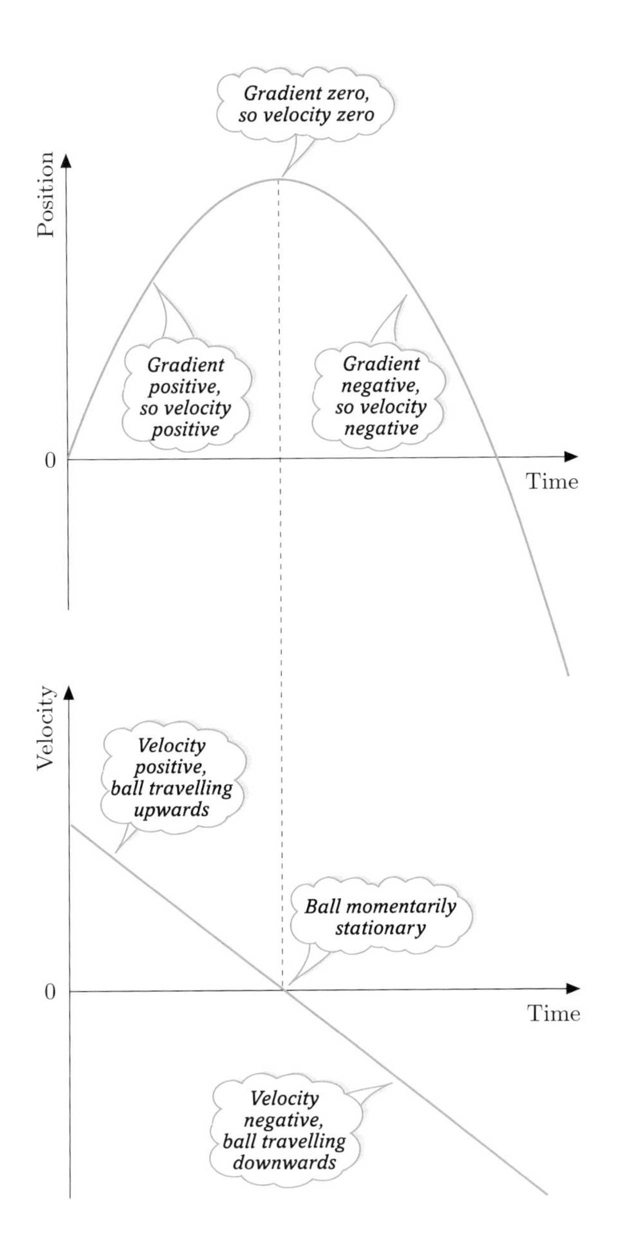

Figure 14

The Earth's gravitational field causes objects thrown into the air, or dropped from a window say, to accelerate at roughly the same constant rate. The constant-acceleration model does not take into account other forces, such as air resistance. So the behaviour of a feather dropped from a window does not fit this constant-acceleration model well, because the air resistance acting upon it is significant. However, close to the surface of the Earth, assuming constant acceleration due to the force of gravity has been found to produce a good model for predicting the motion of objects which do not experience significant air resistance. Galileo is supposed to have tested this model by dropping two cannon balls, one large and one small, from the leaning tower of Pisa and observing that they landed together.

In many situations where acceleration varies considerably, the constant–acceleration model may not be so good, but a model based upon taking the average acceleration (which is constant) may still be useful.

Sometimes, a quantitative model of acceleration is required. In the same way that the slope of a position–time graph gives the velocity, the slope of the velocity–time graph gives the acceleration. So numerical values for acceleration can be worked out from position–time graphs via velocity–time graphs.

Example 2 Acceleration of a ball

Suppose the ball in Activity 2 reaches its highest point after 0.5 seconds and returns to hand level after 1 second, as shown in Figure 15. Find the velocity of the ball at the origin (0 seconds), after 0.5 seconds and after 1 second. Hence calculate the ball's acceleration over its flight, assuming it is constant.

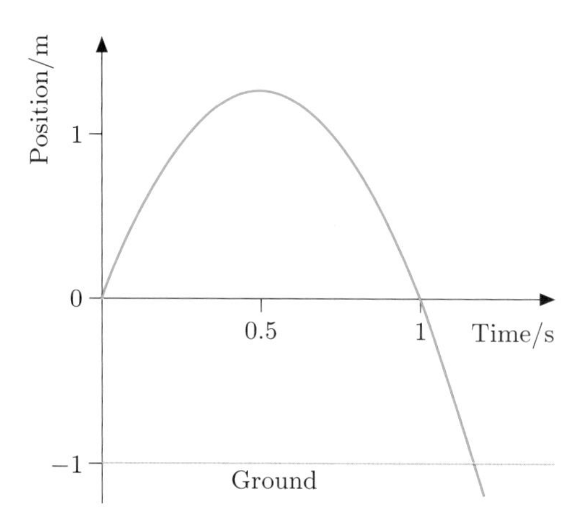

Figure 15

At the origin, the gradient is 1 metre/0.2 seconds $= 5\,\text{ms}^{-1}$ (see Figure 16). This gives the initial velocity with which the ball is thrown in the upward (positive) direction.

After half a second, the ball reaches its highest point and the tangent is horizontal. Hence the gradient is zero, so the velocity is zero there.

A zero gradient occurs at a maximum (or minimum) point, where the graph looks like the top of a hill (or the bottom of a valley).

After 1 second, the ball is back at the same height from which it was thrown (that is, $h = 0$), but the slope of the graph is negative, so the velocity is negative, signalling that the ball is travelling downwards. The gradient here is $1/(0.8 - 1) = -5\,\text{ms}^{-1}$ (see Figure 16), which is the velocity with which the ball passes the point from which it was thrown, on the way down; that is, a speed of $5\,\text{ms}^{-1}$ in a downward direction.

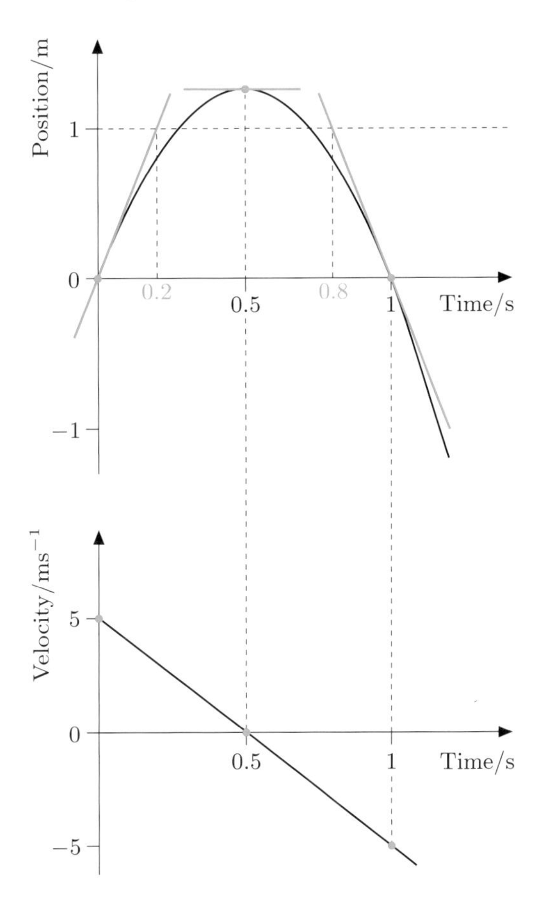

Figure 16

The velocities at these three points can be plotted on the velocity–time graph. These plotted points can then be joined by a straight line, as in Figure 16, since a constant rate of change of velocity—in other words, constant acceleration—is assumed. The constant gradient of this velocity–time graph is the acceleration.

'Metres per second per second' may sound like an odd unit, but it tells you how much faster (or slower if it is negative) you are going at the end of one second: you might care to think of it as '(metres per second) per second'.

The gradient is:

$$\frac{(0-5)\text{ metres per second}}{0.5\text{ second}} = \frac{0-5}{0.5}\text{ metres per second per second}$$
$$= -10\,\text{ms}^{-2}$$

This value is the model's prediction for the acceleration due to gravity at the surface of the Earth, correct to one significant figure, based on the data implicit in Figure 15. The minus sign indicates the acceleration is in a downward direction.

The calculation used only two of the three data points. You might like to check that you obtain the same value for the gradient no matter which two data points you use.

The assumption of constant acceleration leads to a straight-line velocity–time graph—that is, a linear model for velocity—which in turn leads to a curve like that in Figure 15 for the position–time graph. This is the graph of a quadratic function, the details of which are covered in the next section. A straight-line velocity–time graph always leads to a quadratic position–time graph, and vice versa. For the purposes of this course, you need to know, but not prove, that the rate of change of a quadratic function is a linear function.

Calculus

You have seen in earlier units that, although graphs can be drawn from empirical data, another powerful way of generating a graph is from a mathematical function. Many models propose particular functions that fit the data. If you have a function which fits a position–time graph, there is a systematic method of obtaining the corresponding function which describes its slope (the velocity–time function), and vice versa. This involves a branch of mathematics called *calculus*. Going from a position–time function to the corresponding velocity–time function invokes a process called *differentiation*. The velocity–time function, which arises as the rate of change of the position–time function, is called the *derivative* of this latter function. The reverse process to differentiation is called *integration*. It is interesting that the two words, differentiation and integration, have rather different meanings in mathematics from their meanings in everyday life.

Activity 7 Describe the motion

Look at the position–time graph in Figure 17, which shows the position of a train relative to the first station on its route. Sketch the corresponding velocity–time graph. Describe the different sections OA, AB, BC and CD of the journey in terms of the velocity and acceleration of the train.

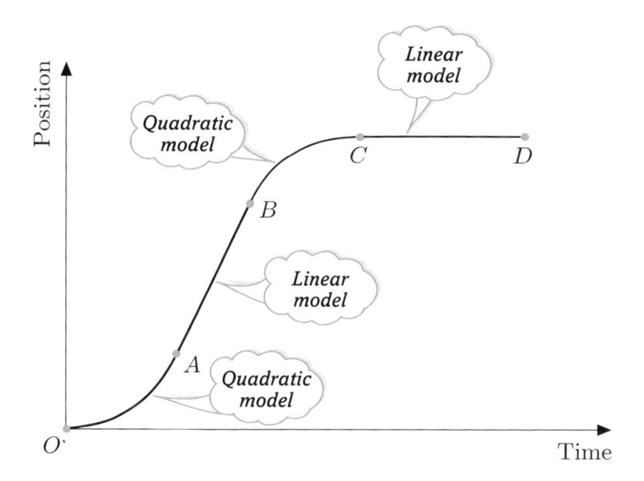

Figure 17

Activity 8

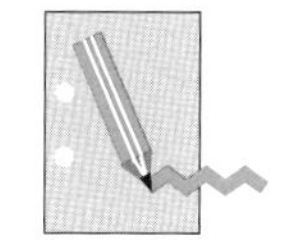

Describe, as if to a friend, how velocity and acceleration on a journey are related to the position–time graph of the journey. Use your answer to help you refine the meanings of the new terms you have included on the Handbook sheet.

In summary, this section has introduced some of the mathematical language that can be used to describe or model situations involving motion. In such models, some aspects are stressed, such as directions and rates of change; others are ignored, such as the size and shape of the moving object and any deviation from one-dimensional movement.

When describing the motion of something, it is important to specify the starting point (origin) from which all measurements are made, and which direction is taken as positive. The opposite direction is then considered to be the negative direction.

Velocity is the rate of change of position and is given by the gradient of the position–time graph. It is an extension of the idea of speed, but differs from the notion of speed in that the direction of travel is also important: a positive gradient on a position–time graph represents a positive velocity, indicating that the motion is in the positive direction; a negative gradient on a position–time graph represents a negative velocity, indicating that the motion is in the negative direction.

Acceleration is the rate of change of velocity, and is given by the gradient of the velocity–time graph. An object is accelerating whenever its velocity changes, whether it is speeding up or slowing down.

Constant-acceleration models are often useful, for example for modelling falling objects. They lead to a linear function for velocity against time and a quadratic function for position against time.

Outcomes

After studying this section, you should be able to:

◇ use correctly the terms 'position', 'velocity' and 'acceleration' in describing the motion of something (Activities 1, 5, 6, 7);

◇ use a position–time graph to sketch the corresponding velocity–time graph, and explain how velocity and acceleration are related to the gradients of such graphs (Activities 1, 2, 3, 4, 5, 7, 8).

2 Quadratic models

Aims This section aims to introduce the quadratic function to your library of standard functions, along with its associated graph: the parabola. ◇

In some situations involving motion, using a linear function is inappropriate: for instance, for describing the position of an accelerating vehicle. As you saw in Section 1, when modelling such motion, a curve rather than a straight line provides a more realistic position–time graph. One of the simplest curves is the parabola, which arises as the graph of a quadratic function. This section first looks at the mathematics of quadratic functions and the properties of parabolas, before returning to situations where they are useful modelling tools. It also looks at the fitting of quadratic functions to data using regression.

2.1 Quadratic functions and parabolas

Figure 18 (overleaf) shows some parabolas. A *parabola* is a symmetrical curve; if you imagine a parabola folding along its own 'centre line', called its *axis*, the two halves would lie exactly on top of each other. The point where the parabola cuts its axis is called the *vertex*.

When the axis of a parabola is parallel to the y-axis, as in Figures 18(a) and 18(b), the vertex is the highest or lowest point of the parabola. As you can imagine, a curve with a highest or lowest point like this is very useful in modelling: for instance, a curve like Figure 18(a) could be used to model the motion of a ball thrown directly up into the air; a curve like Figure 18(b) could be used to model the speed of a vehicle as it passes an obstruction. In fact, if you are modelling a situation where one quantity has either a highest or a lowest point, generally a useful first step is to assume that this situation can be described by a parabola.

(Other curves also have highest and lowest points, but in general they are not as easy to handle mathematically as parabolas are, so it is usually worth trying a parabola first.)

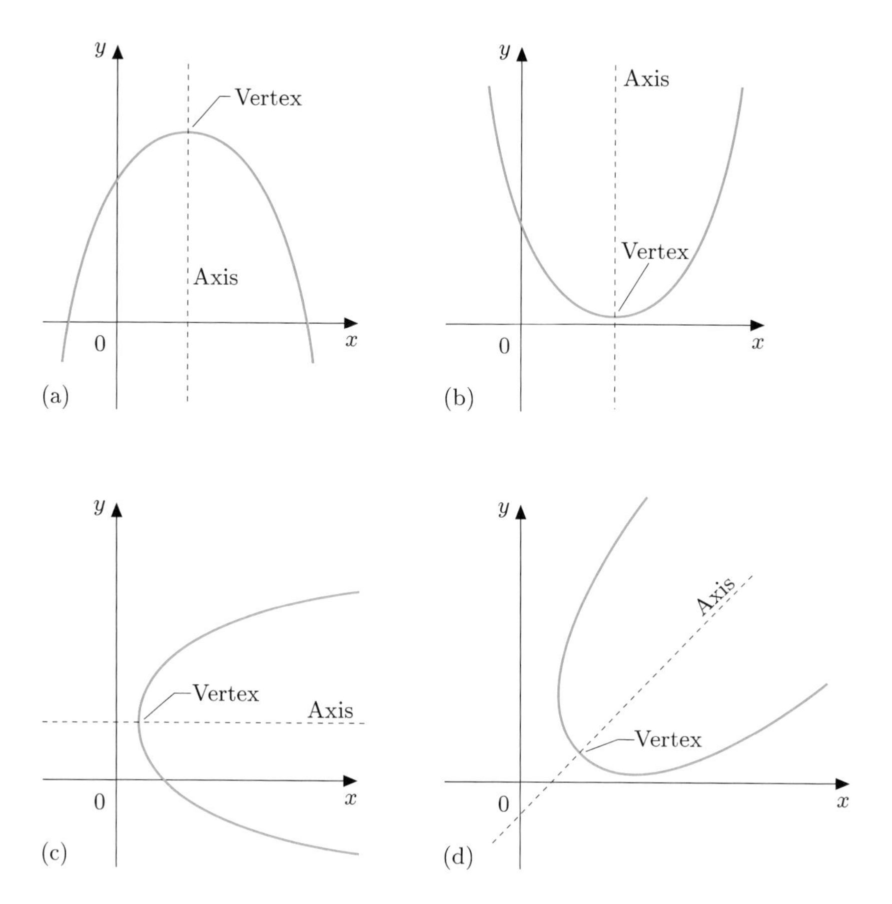

Figure 18 Some parabolas

In Figure 18(c), the axis of the parabola is parallel to the x-axis, while Figure 18(d) shows a parabola where its axis is at an angle to both the x- and y-axes. Such curves are less useful in the sort of modelling covered by this course, and also are more difficult to plot on your calculator. However, both sorts of parabola can be transformed into ones with axis parallel to the y-axis—though the details of this are beyond the scope of this course. This course only considers parabolas whose axes are parallel to the y-axis (and hence which have either a highest or lowest point).

Figure 19 shows four parabolas in what is known as *standard position*, with their vertices at the origin and their axes along the y-axis. The formula or equation specifying such a parabola is

$$y = ax^2$$

The quantity a is known as a *parameter* of the equation.

where a can be any number. If a is positive, then the parabola opens upwards; if a is negative, it opens downwards. The smaller the value of a, the more 'open' is the parabola; the larger the value of a, the more 'closed' is the parabola.

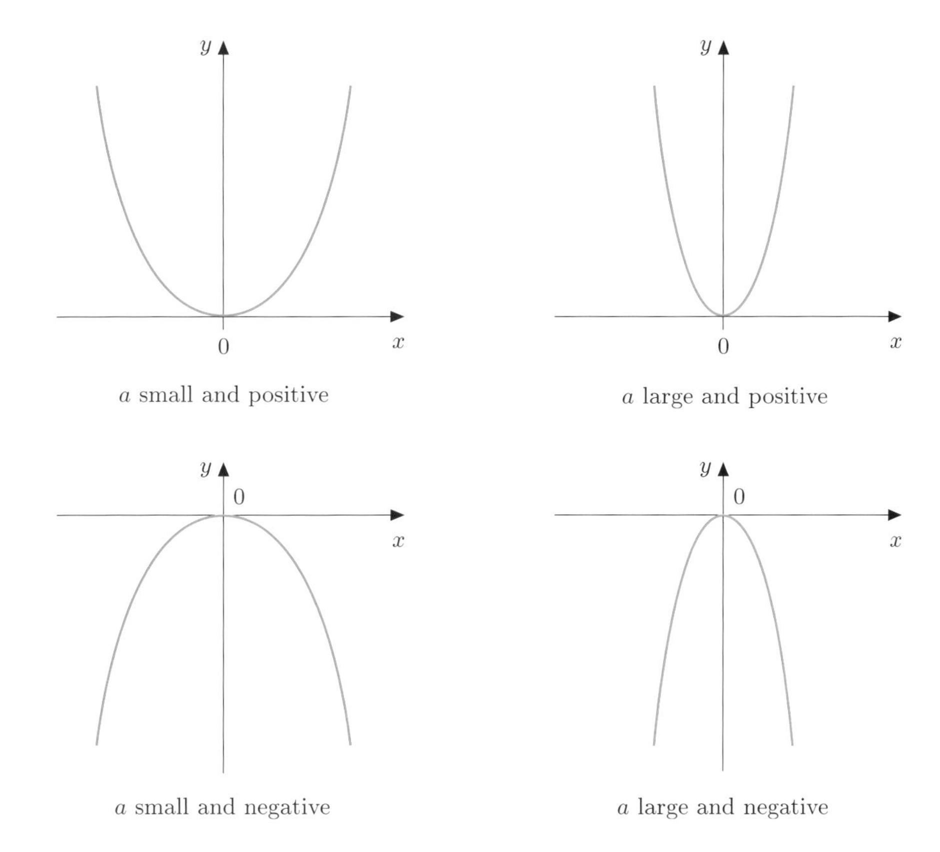

Figure 19 Parabolas in standard position

Activity 9 *Plotting a parabola*

On your calculator, plot the graphs of the following quadratic functions.

(a) $y = x^2$
$y = 4x^2$
$y = 9x^2$

(b) $y = -x^2$
$y = -4x^2$
$y = -9x^2$

In your own words, explain how the graph of a quadratic function $y = ax^2$ depends upon the numerical value of the parameter a.

Plotting graphs for $y = ax^2$ for different values of the parameter a produces a family of parabolas which all pass through the origin $(0, 0)$. The numerical size of the parameter a determines whether the parabola is narrow or wide: the smaller a is, the wider is the parabola.

There are two parameters in addition to a that are involved in the general specification of a quadratic function. In both cases, the graph of the quadratic function is a parabola whose vertex is not necessarily at the origin. You are asked to investigate the effect of the first of the parameters in the following activity.

Activity 10 Moving a parabola up or down

On your calculator, plot the graphs of the following quadratic functions. In each case, note the position of the vertex of the parabola.

(a) $y = x^2$
$y = x^2 + 1$
$y = x^2 + 2$
$y = x^2 - 1$

(b) $y = -2x^2$
$y = -2x^2 + 1$
$y = -2x^2 + 2$
$y = -2x^2 - 1$

In your own words, explain how the graph of a quadratic function $y = ax^2 + l$ depends upon the numerical value of the parameter l.

So now you can move a parabola up or down by adding, respectively, a positive or negative number l to the function $y = ax^2$ to get $y = ax^2 + l$. The family of graphs with equations of the form $y = ax^2 + l$ are parabolas with their vertices on the y-axis at $(0, l)$.

► You can now move a parabola up or down, but how could you move a parabola sideways?

There are a number of ways to think of this. Here is one which you might find helpful. (If you find another way which works for you, do use it and perhaps explain it to your fellow students and tutor, or write a description in your Learning File.) $y = ax^2 + l$ can be rearranged as:

$$(y - l) = ax^2$$

So the equation $y = ax^2 + l$ is similar to the standard position equation $y = ax^2$ but with $(y - l)$ instead of y. Look back at your description of how the graph depends upon the value of l. The vertex of the parabola moves up from the origin by l units, to the point $(0, l)$; so replacing y by $(y - l)$ moves the vertex l units along the y-axis.

Therefore, to move the graph k units along the x-axis, it makes sense to try replacing x by $(x - k)$ in the standard position equation $y = ax^2$.

Activity 11 Moving a parabola sideways

Confirm that replacing x by $(x - k)$ in the standard position equation of a parabola moves the vertex k units along the x-axis, by plotting the graphs of the following functions on your calculator. In each case, note the position of the vertex of the parabola.

(a) $y = x^2$
$y = (x - 1)^2$
$y = (x - 2)^2$
$y = (x + 1)^2$

(b) $y = -2x^2$
$y = -2(x + 1)^2$
$y = -2(x + 2)^2$
$y = -2(x - 1)^2$

In your own words, explain how the graph of a quadratic function $y = a(x - k)^2$ depends upon the numerical value of the parameter k.

So you can now move a parabola up or down, or to the left or right.

▶ How could you do both: that is, move the vertex of a parabola to any position?

You need to replace y by $(y - l)$ and x by $(x - k)$ in the standard position equation $y = ax^2$.

For instance, if you want the y-coordinate of the vertex to be 2, you need to replace, y by $(y - 2)$, and if you want the x-coordinate of the vertex to be 4, replace x by $(x - 4)$. So the general equation producing all parabolas whose vertices are at $(4, 2)$ is $(y - 2) = a(x - 4)^2$. In order to write this expression in the form of a function for your calculator, you need to rearrange it by adding 2 to both sides, to give $y = a(x - 4)^2 + 2$. To get a particular graph from your calculator, you need to choose a particular numerical value for a.

Activity 12 Moving a parabola in any direction

On your calculator, plot the graphs of the following.

(a) $y = (x - 4)^2 + 2$

(b) $y = 2(x - 4)^2 + 2$

(c) $y = -2(x - 4)^2 + 2$

(d) $y = 3(x - 4)^2 + 2$

You should now have verified that the family of parabolas with vertices at $(4, 2)$ have equations of the form:

$$y = a(x - 4)^2 + 2$$

In general, if the vertex of a parabola is at a point (k, l), then its equation is of the form:

$$(y - l) = a(x - k)^2$$

or equivalently:

$$y = a(x - k)^2 + l \qquad (1)$$

This fact is useful when modelling with quadratic functions, where you know the vertex of the parabola: you can just substitute in the coordinates of the vertex for k and l in equation (1).

For example, any parabola with its vertex at the point $(-1, 5)$ and its axis vertical has an equation of the form:

$$y = a(x - (-1))^2 + 5$$

which simplifies to:

$$y = a(x + 1)^2 + 5 \qquad (2)$$

A particular value for a can be determined by knowing at least one other point that lies on the curve. So, for example, if you know that the curve also passes through the origin $(0, 0)$, then the values $x = 0$, $y = 0$ must satisfy equation (2). Substituting these values in equation (2) gives:

$$0 = a(0 + 1)^2 + 5$$

So $a = -5$, and the equation of the particular parabola is:

$$y = -5(x + 1)^2 + 5$$

Activity 13

(a) Write down the general equation (that is, one with a in) of any parabola whose axis is parallel to the y-axis and whose vertex is at:

(i) $(3, 0)$; (ii) $(-2, -1)$.

Check your equations by plotting the graphs on your calculator with $a = 1$.

(b) What are the coordinates of the vertex of each of the following families of parabolas?

(i) $y + 3 = a(x - 9)^2$ (ii) $y = 2 + a(x + 6)^2$

Often, in modelling, the formula for a parabola does not arise in the form given in equation (1). Instead, it occurs in a multiplied-out form. To illustrate this, look at equation (2) for the particular case when $a = 1$:

$$y = (x + 1)^2 + 5 \qquad (3)$$

As you saw in *Unit 8*, $(x + 1)^2$, which is the same as $(x + 1)(x + 1)$, can be multiplied out as follows:

$$\begin{aligned}(x+1)(x+1) &= x(x+1) + 1(x+1)\\ &= x^2 + x + x + 1\\ &= x^2 + 2x + 1\end{aligned}$$

So equation (3) becomes:

$$y = (x^2 + 2x + 1) + 5$$

which simplifies to:

$$y = x^2 + 2x + 6 \qquad (4)$$

▶ Plot the two functions in equations (3) and (4) on your calculator to check that they give the same graph.

Both equations produce the same parabola, but in form the equations are dissimilar.

In general, when you draw the graph of any function of the form

$$y = ax^2 + bx + c \qquad (5)$$

(where the parameters a, b and c are any numbers, except that a must not be zero), you get a parabola. Equation (5) is an alternative form of equation (1)—as is demonstrated algebraically in Section 1 of the Appendix. The sign of a determines which way up the parabola is. The value of c gives the y-intercept of the curve, just as for a straight-line graph. An algebraic argument (also given in Section 1 of the Appendix) shows that the vertex is at:

$$\left(-\frac{b}{2a},\ c - \frac{b^2}{4a}\right)$$

Example 3 *Is it a parabola?*

Does the function

$$y = 5x^2 + 40x + 81$$

produce a parabola? If so, what are the coordinates of the vertex?

Yes. This fits the form of equation (5) with $a = 5$, $b = 40$ and $c = 81$. So its graph is a parabola.

The vertex coordinates are:

$$-\frac{b}{2a} = \frac{40}{2 \times 5} = -4 \qquad \text{(for the } x\text{-coordinate)}$$

$$c - \frac{b^2}{4a} = 81 - \frac{40^2}{4 \times 5} = 81 - 80 = 1 \qquad \text{(for the } y\text{-coordinate)}$$

So the vertex is at $(-4, 1)$.

Check: $y = 5(x-(-4))^2 + 1$ should be the same as $y = 5x^2 + 40x + 81$. This is so because:

$$\begin{aligned} 5(x-(-4))^2 + 1 &= 5(x+4)(x+4) + 1 \\ &= 5(x^2 + 8x + 16) + 1 \\ &= (5x^2 + 40x + 80) + 1 \\ &= 5x^2 + 40x + 81 \end{aligned}$$

The general function in equation (5), when a is non-zero, defines a *quadratic function*: that is, any quadratic function can be written as

$$y = ax^2 + bx + c, \qquad \text{where } a \text{ is non-zero.}$$

Rate of change of a quadratic function

Earlier, you met the idea that the rate of change of a quadratic function, which is the same as the slope of its graph, is a linear function, and that this function is called the derivative. The *Calculator Book* has an investigation to help convince you of this, which you will be asked to look at shortly.

In fact, the derivative of the quadratic function

$$y = ax^2 + bx + c$$

is the linear function

$$y = 2ax + b$$

However, you do not need to remember this for MU120.

At the vertex of a parabola whose axis is parallel to the y-axis, the slope of the curve is zero; so the derivative is zero there. This is true for the maximum and minimum points of other functions too.

2.2 Fitting a parabola to data

In order to fit a specific straight line to data, you need at least two points. In order to fit a parabola, you generally need at least three points. However, if one point is actually the vertex, you only need two points. This is because, as you saw earlier, you can substitute the coordinates of the vertex for k and l into equation (1) and then substitute the coordinates of the other point into the resulting equation to find the value of a. In any other situation, where you have three or more data points to which you wish to fit a parabola, you can use your calculator to do so, in a similar way to when fitting a straight line. However, this time you need to use *quadratic regression* instead of linear regression.

Now work through Sections 11.1–11.3 of Chapter 11 of the Calculator Book.

Activity 14 Fitting a quadratic function to throwing a ball

Suppose that you want to fit a quadratic function to the motion of the ball in Figure 15. The data points you have are as follows.

Time (seconds)	0	0.5	1
Height above hand (metres)	0	1.25	0

Use your calculator to fit a quadratic function to these data points, and use it to predict when the ball will hit the ground (a height of 1 metre *below* the hand).

Activity 15 Summarizing ideas

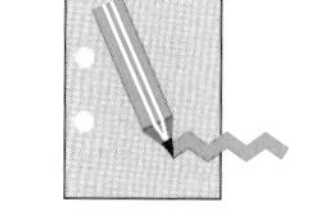

Summarize, in your own words, the main properties of quadratic functions and parabolas for your Handbook.

2.3 Using quadratic models

This subsection considers the use of *quadratic models*—that is, models involving a quadratic function or a parabola.

Activity 16 A dummy's head

Suppose you have data from the frames of a video of a simulated accident involving a car, with a dummy in the driving seat, crashing into a solid object. The data give the position of the dummy's head relative to the windscreen at different times after the impact.

Time (seconds)	0	0.01	0.02	0.03	0.04	0.05	0.06	0.07	0.08
Position (metres)	1.00	0.97	0.93	0.87	0.75	0.61	0.43	0.24	0.00

(a) Use your calculator to fit a quadratic function to this data, using quadratic regression.

(b) Use algebra to fit a quadratic function, assuming that $(0, 1)$ is the vertex and $(0.08, 0)$ is a point on the curve.

(c) Enter this function into your calculator alongside the quadratic regression function. Use the table facility of your calculator to see how good a fit each function is to the data points.

(d) Plot both functions and the data points on your calculator, and comment on the two models that the functions embody.

(e) Use the functions and your calculator to estimate the speed with which the dummy's head hits the windscreen.

Quadratic functions may be useful in contexts other than motion. An example is modelling the population density of a town or city. Often population densities are low in city centres, as most buildings are used for commercial rather than residential purposes. Surrounding the centre are densely populated residential areas. Further out are the more spacious suburbs, where population density is again lower. An algebraic model for population density can be useful for analysis of trends and comparisons, as well as for prediction—both extrapolation and interpolation. An algebraic model also allows parameters to be changed easily when the model is updated. Population density models were used in the analysis of asthma clusters, shown in the *Asthma and the Bean* television programme, and they are also useful in the design and planning of local services and businesses.

Activity 17 Population density

Suppose that aerial photography indicated that, for a particular town, the population density mostly depends on the distance from the centre, with a peak population density 2 km from the centre. Data from a land-based survey were available for two areas, indicating that the population density was about 6000 people per km^2 at the town centre and 30 000 people per km^2 at a distance of 2 km from the centre.

(a) Find a quadratic function that could be used to model the variation of population density with distance from the town centre. Use it to predict the population density 0.5 km from the centre. (Use quadratic regression on your calculator or an algebraic method, whichever you prefer.)

(b) Use your calculator to draw the parabola corresponding to the quadratic function you found in (a), and check that its vertex corresponds to a population density of 30 000 people per km^2 at 2 km from the centre and that its intercept corresponds to a population density of 6000 people per km^2 at the centre.

(c) For what range of distances from the centre might your model be useful?

Activity 18 CD sales

A CD company sells 10 000 copies of a pop CD in the week that the CD is released, rising to 22 000 in the sixth week of sales. After that, the sales begin to fall off. So that they can plan their output, they want to predict future sales every week. Find a simple quadratic function that will enable them to do this. Use this function and your calculator to produce a table of expected sales from the sixth week of sales onwards. When does your model suggest that sales will fall to zero?

So far in this section, the letters x and y have been used as names for the variables. However, as is the case with linear models, it is very often more sensible to use other letters for the variables in quadratic models: for example, s for speed or t for time. (The letter s is also commonly used for position or distance.)

Be careful not to confuse the use of the italic s for speed, position or distance with the use of the non-italic s for seconds.

Activity 19 Goods train

Suppose that you are trying to schedule an additional goods train into an existing timetable. Your information suggests that the goods train will travel a distance s km in t hours, where $s = 80t^2$ $(0 \leq t \leq 0.5)$. Plot this function, and produce a table of distances travelled every 10 minutes over a 30-minute period.

To summarize, this section has explained the idea of a parabola and how parabolas can be represented by quadratic functions. It has also shown how quadratic regression can be used to fit a parabola to data, and has provided examples of the use of quadratic models.

Outcomes

After studying this section, you should be able to:

- ◇ explain the significance of the parameters a, b, c, k and l in the two general expressions for quadratic functions $y = a(x - k)^2 + l$ and $y = ax^2 + bx + c$ (Activities 9, 10, 11, 12, 13, 15);
- ◇ find the vertex of a parabola (Activity 13);
- ◇ fit a quadratic function to appropriate given data (Activities 14, 16, 17, 18);
- ◇ use a quadratic function to make predictions (Activities 14, 16, 17, 18, 19);
- ◇ find the gradient at a given point of a parabola (Activity 16).

3 Quadratic equations

Aims This section builds upon the calculator work of the previous section, in order to find where a parabola crosses the x-axis. It aims to teach you how to find such solutions algebraically as well as graphically, so that you can solve quadratic equations involving parameters when they arise in modelling. ◇

On occasion, you may find you need to know when a function takes a particular value. You can use the trace or table facilities on your calculator, but sometimes an algebraic solution is preferable—for example, for an equation containing parameters. When the function is linear, as in *Unit 10*, you can use algebraic manipulation to solve the equation. However, algebraic methods are less straightforward for quadratic functions. Linear equations always have one numerical solution; but, because quadratic functions 'bend over', quadratic equations may have two numerical solutions or no solution at all. There are numerous methods for solving quadratic equations and this section covers several of them. Sometimes one method is preferable, but often there is a choice of method.

3.1 Quadratic equations and their solutions

Suppose that, as in Activity 19, you have modelled the motion of a goods train pulling out of a station by the quadratic function $s = 80t^2$, where s is the distance travelled in km and t is the time in hours ($0 \leq t \leq 0.5$). Rather than produce a table of distances, you might simply wish to predict when the train should have travelled 5 km, to a crucial junction perhaps; that is, what is the value of t when s is 5? In order to find this, you need to solve the following equation:

$$5 = 80t^2 \qquad (6)$$

This is an example of a *quadratic equation*. A quadratic equation is an equation in which a quadratic function equals a constant. The previous section talked about equations of parabolas. These were formulas, relating y to x, and the parabola was the shape of the graph generated by the function given by the formula. Here, the term 'quadratic equation' will have the particular meaning 'quadratic function equals constant'.

The most straightforward algebraic method for solving quadratic equations can be illustrated using equation (6). First, divide through by 80 to get t^2 on its own, giving:

$$\tfrac{1}{16} = t^2$$

Then, taking square roots of both sides (noting that t represents time after the train left the station, so is positive, and so only the positive square root need be considered), gives:

$$\tfrac{1}{4} = t$$

So the model predicts that the train will reach the 5 km junction after a quarter of an hour.

Activity 20 When will the ball reach this height?

Consider a more complicated quadratic function, modelling a ball thrown vertically in the air. The height h (in metres) of the ball above its starting point at a time t seconds after being thrown is given as:

$$h = 5t - 5t^2 \qquad (0 \leq t \leq 1.2) \tag{7}$$

(a) Plot this graph on your calculator and use the trace facility to find when (correct to two decimal places) the ball is 1 metre above its starting point. You should find *two* such points—one on the way up and the other on the way down.

(b) Substitute each of the values that you found for t into equation (7) and check that they both give the value of h as approximately 1 (that is, that the ball is at a height of 1 metre at these times, as required).

(c) Suppose that you wanted to predict when the ball reached other heights. Use your calculator and the trace facility to find any values of t for which h is 1.25 and 1.5.

The above activity involved solving equation (7) for $h = 1, 1.25, 1.5$. In other words, you were solving the following quadratic equations:

$$\begin{aligned} 1 &= 5t - 5t^2 \\ 1.25 &= 5t - 5t^2 \\ 1.5 &= 5t - 5t^2 \end{aligned}$$

In the first case, there were two solutions; in the second, there was only one solution; in the third case, there was no solution at all.

▶ Why is this so? Look at Figure 20 and try to see why.

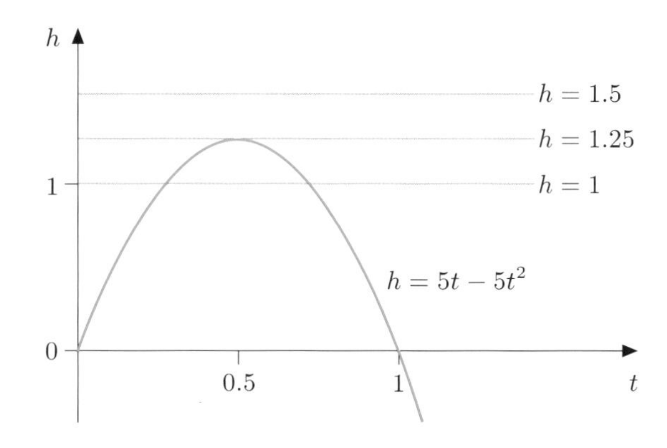

Figure 20

A graphical way of thinking about solving a quadratic equation is to look for the intersection of a horizontal line like $h = 1$, $h = 1.25$ or $h = 1.5$ and the parabola that is the graph of the corresponding quadratic function. The line may cut the parabola twice, as for $h = 1$, in which case there are two solutions. The line may just touch the parabola at one point (the vertex), as for $h = 1.25$, in which case there is one solution. Alternatively, the line may not cut the parabola at all, as for $h = 1.5$, and then there is no solution.

3.2 General methods of solving quadratic equations

You may have met quadratic equations before and may know ways of solving them already. If this is the case, try Activities 25, 26 and 27 now. If you have no problems with them, skip the rest of this subsection and go straight to Section 4. If you do not how to solve quadratic equations, or if you have difficulties with Activities 25, 26 and 27, study this subsection in full.

Method 1: graphical solution

Quadratic equations can be solved by graphical means. For instance, suppose you want to use the model from Activity 17 to predict where the population density is 10 000 people per km^2: that is, where the line $y = 10\,000$ cuts the graph of $y = -6000x^2 + 24\,000x + 6000$. Figure 21 shows that this is at approximately $x = 0.2$ and $x = 3.8$. The solutions of the quadratic equation

$$10\,000 = -6000x^2 + 24\,000x + 6000$$

or equivalently

$$0 = -6000x^2 + 24\,000x - 4000$$

are therefore approximately $x = 0.2$ and $x = 3.8$.

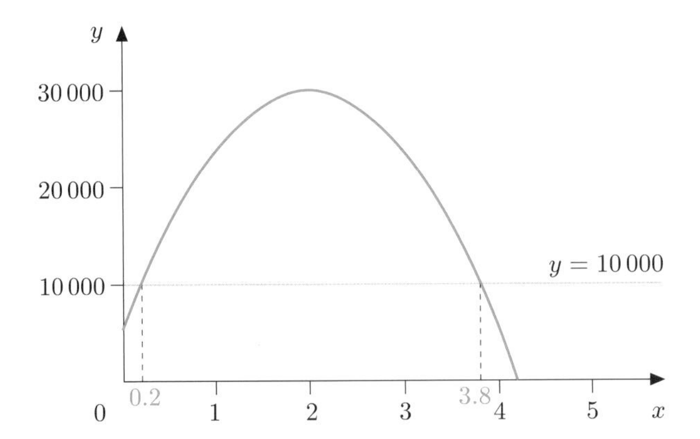

Figure 21

So the population density is likely to be 10 000 per km^2 at 0.2 km from the town centre and again at 3.8 km out.

Activity 21

You can get more accurate values for the distances by graphing the function $y = -6000x^2 + 24\,000x - 4000$ on your calculator and using the trace and zoom facilities to find the values of x where the graph crosses the x-axis (that is, where $y = 0$). Do this now and so predict these distances to two decimal places.

Method 2: using the table facility on the course calculator

This is often quicker than zooming, once you have found an approximate solution, for example by using method 1. Set the table minimum to value close to (but less than) the approximate solution and set the table increment to an appropriate level of accuracy, and then move down to the nearest value to the solution.

▶ Try this instead of zooming for Activity 21. Do you prefer it?

Method 3: finding the square root

The simplest types of quadratic equations are the ones of the form $ax^2 = g$, which do not contain any term of the form bx. Such equations can be solved simply by taking square roots. For example, consider the following quadratic equation:

$$3x^2 = 8$$

Dividing both sides by 3 gives:

$$x^2 = \frac{8}{3}$$

So, taking the square root of both sides of the equation gives:

$$\begin{aligned} x &= +\sqrt{\frac{8}{3}} \text{ or } -\sqrt{\frac{8}{3}} \\ &= +1.63 \text{ or } -1.63 \qquad \text{(to two decimal places)} \end{aligned}$$

A piece of mathematical shorthand which is useful in writing pairs of solutions like this is to write:

$$x = \pm 1.63$$

This means 'x equals either plus 1.63 or minus 1.63' and saves writing out both the positive and negative values explicitly.

In general, the solution of an equation of the form

$$ax^2 = g$$

where a and g are constants, and a is non-zero, is:

$$x = \pm\sqrt{\frac{g}{a}}$$

Notice that in order to have a solution, g/a must be positive (or zero), because no ordinary number has a negative square. So if g/a is negative, then there is no solution.

Another type of quadratic equation is exemplified by the following:

$$(x+5)^2 = 9$$

Here, taking the square root gives either

$$x+5=+3 \quad \text{or} \quad x+5=-3$$

which may be written using the symbol $\pm$ as follows:

$$x+5=\pm 3$$

And so the solutions are:

$$x=-5\pm 3$$

that is:

$$x=-5+3 \quad \text{and} \quad x=-5-3$$

or

$$x=-2 \quad \text{and} \quad x=-8$$

In general, the solutions of an equation of the form:

$$(x+m)^2 = n$$

where m and n are constants, are found by taking square roots, to give:

$$x+m=\pm\sqrt{n}$$

and therefore:

$$x=-m\pm\sqrt{n}$$

Again, notice that in order to have a solution, n cannot be negative, because no 'ordinary' number has a negative square. If n is negative, then there is no solution to the quadratic equation, as there is no 'ordinary' number $x+m$ whose square is a negative number.

Imaginary numbers

A branch of mathematics, called *complex analysis*, is based on the idea of the square root of a negative number. Such a square root is called an *imaginary number*, and the 'ordinary' numbers you are accustomed to are called *real numbers*.

Activity 22

Find the real-number solutions (if any) of the following quadratic equations.

(a) $x^2 = 25$

(b) $9x^2 = 16$

(c) $(y + 1)^2 = 16$

(d) $(x - 3)^2 = 15$

(e) $(z + 2)^2 + 9 = 0$

Check your solutions by substituting them in the relevant equation.

Method 4: factorization

A different type of quadratic equation is one where there is no constant term, such as the following:

$$2x^2 + 5x = 0$$

Here x is a *common factor* to both terms on the left-hand side, and so the equation can be rewritten as follows:

$$x(2x + 5) = 0$$

Now, in normal arithmetic, if two quantities multiply together to give zero, then *one or other of these two quantities must itself be zero.* So here, either $x = 0$ or $2x + 5 = 0$. This leads to the following two possible solutions:

Note this is a very special property of the number zero alone. If you had to solve $x(2x + 5) = 8$, you cannot assume $x = 8$ or $2x + 5 = 8$, because there are many different ways in which two numbers can multiply together to give 8 (or any other number other than zero for that matter).

$$x = 0 \quad \text{or} \quad x = -2.5$$

The latter is the solution of $2x + 5 = 0$.

In general, the solution of an equation of the form

$$ax^2 + bx = 0$$

where a and b are constants, and a is non-zero, can be found by using the fact that x is a common factor of both terms of the equation. So the equation can be rewritten as $x(ax + b) = 0$. So either $x = 0$ or $ax + b = 0$ (so $ax = -b$ and hence $x = -b/a$). The solutions are thus:

$$x = 0 \quad \text{and} \quad x = -\frac{b}{a}$$

Another type of quadratic equation is exemplified by the following:

$$2x^2 - x - 21 = 0$$

which can be rewritten as:

$$(x + 3)(2x - 7) = 0$$

Multiply out the brackets to check.

This can be solved by noting that if two quantities multiply together to give zero then either one or other of those quantities *must* be zero.

So either

$x + 3 = 0$, in which case $x = -3$;

or

$2x - 7 = 0$, in which case $x = \frac{7}{2}$.

In general, an equation that can be written in the form

$$(lx + m)(nx + p) = 0 \tag{8}$$

where l, m, n and p are constants, and l and n are non-zero, has solutions:

$$x = -\frac{m}{l} \quad \text{and} \quad x = -\frac{p}{n}$$

Many (but not all) quadratic equations can be *factorized*—that is, written in the same form as equation (8). It is not an objective of this course to teach you to factorize quadratic equations, but there is a summary of a technique in Section 2 of the Appendix.

Activity 23

Find the solutions of the following quadratic equations.

(a) $x^2 + 12x = 0$

(b) $3y^2 - 9y = 0$

(c) $5z^2 + z = 0$

(d) $(x - 9)(2x + 11) = 0$

Method 5: the formula method

This is the method which *always* produces the solution(s) of a quadratic equation (if there are any), no matter how 'awkward' the numbers involved are. The formula which gives the solution of the quadratic equation

$$ax^2 + bx + c = 0$$

where a, b and c are constants, and a is non-zero, is:

$$x = \frac{-b \pm \sqrt{(b^2 - 4ac)}}{2a}$$

The derivation of the formula is given in Section 3 of the Appendix.

Example 4 Using the formula method

Use the formula method to solve the following quadratic equation:

$$3x^2 + x - 6 = 0$$

Here $a = 3$, $b = 1$ and $c = -6$. (Note the need to include the minus sign in the constant term: using $c = 6$ would be incorrect.)

In general, $x = \dfrac{-b \pm \sqrt{(b^2 - 4ac)}}{2a}$. So here:

$$\begin{aligned} x &= \frac{-1 \pm \sqrt{[1^2 - 4 \times 3 \times (-6)]}}{2 \times 3} \\ &= \frac{-1 \pm \sqrt{(1 + 72)}}{6} \\ &\simeq \frac{-1 \pm 8.544}{6} \\ &= 1.26 \text{ or } -1.59 \qquad \text{(to two decimal places)} \end{aligned}$$

So the solutions are:

$x = 1.26$ and $x = -1.59$ (to two decimal places)

Section 11.4 of Chapter 11 the *Calculator Book* discusses a program for speeding up the calculation of solutions using the formula method. The discussion and use of the program are *optional.*

If you wish, now work through Section 11.4 of Chapter 11 of the Calculator Book.

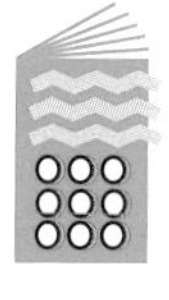

If you are using the formula and the number under the square root sign turns out to be negative, then there is no real-number solution to the equation.

Activity 24

Use the formula method to solve (if possible) the following quadratic equations.

(a) $x^2 - 9x + 2 = 0$

(b) $2y^2 + 3y - 5 = 0$

(c) $4m^2 + 12m + 9 = 0$

(d) $5z^2 - 2z + 2 = 0$

Check your answers.

Notice that the formula gives the three cases mentioned in Subsection 3.1.

If $b^2 > 4ac$ (as in Activity 24(a) and (b)), there are two solutions (one from the + sign and the other from the − sign in front of the square root).

If $b^2 = 4ac$ (as in Activity 24(c)), there is only one solution (because the square-root term is zero).

If $b^2 < 4ac$ (as in Activity 24(d)), then any solution would require the square root of a negative number, so there are no real-number solutions.

Activity 25 Population density

In Activity 17, you arrived at the following formula for modelling the variation of population density with distance from a town centre:

$$y = -6000x^2 + 24\,000x + 6000$$

Suppose you are tendering for certain services which depend upon population densities, and that you want to find the distances from the town centre for which the population density is 10 000 people per km^2 or more. So you are interested in finding the value(s) of x for which $y \geq 10\,000$. In Activity 21, you used your calculator to find graphically the value(s) of x which give $y = 10\,000$, by plotting the graph of the function

$$y = -6000x^2 + 24\,000x - 4000$$

and finding where it crossed the x-axis. These values of x can also be found algebraically by solving the following equation:

$$0 = -6000x^2 + 24\,000x - 4000$$

(a) Use the formula method to solve this equation, giving your answer to two decimal places.

Hint: you can divide the quadratic equation through by a very large common factor before you start to solve it.

(b) Hence predict how many kilometres from the town centre the population density is more than 10 000 people per km^2.

Activity 26

Solve (if possible) the following quadratic equations by whatever method you find most convenient.

(a) $x^2 + 5x + 6 = 0$

(b) $y^2 - 10y + 16 = 0$

(c) $p^2 + 4p + 1 = 0$

(d) $x^2 + x - 1 = 0$

(e) $4y^2 - 9y + 27 = 0$

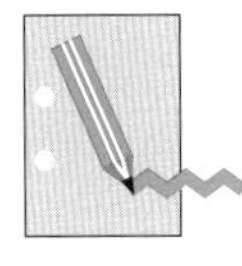

Activity 27 Summarizing ideas

Summarize your preferred method(s) for solving quadratic equations, as if describing them to a friend, and explain the circumstances when there are two, one or no solution(s).

In summary, a quadratic equation is an equation of the form

$$ax^2 + bx + c = 0$$

where a, b and c are parameters with $a \neq 0$. A quadratic equation may have two solutions, one solution or no real-number solutions.

If $b^2 > 4ac$, there are two solutions.
If $b^2 = 4ac$, there is one solution.
If $b^2 < 4ac$, there are no real-number solutions.

A quadratic equation can be solved graphically (with or without the additional use of a table) using the course calculator, or algebraically by several methods.

(a) The formula method gives solutions to $ax^2 + bx + c = 0$ $(a \neq 0)$ by the following formula:

$$x = \frac{-b \pm \sqrt{(b^2 - 4ac)}}{2a}$$

(b) The square-root method gives solutions to

$$ax^2 = g \qquad (a \neq 0, g/a \geq 0)$$

as $x = \pm\sqrt{\dfrac{g}{a}}$;
and solutions to

$$(x + m)^2 = n \qquad (n \geq 0)$$

as $x = -m \pm \sqrt{n}$.

(c) The factorization method finds solutions to the following:

$$ax^2 + bx = 0 \qquad (a \neq 0)$$

This factorizes to

$$x(ax + b) = 0$$

and so has solutions $x = 0$ and $x = -b/a$.

It also gives solutions to

$$(lx + m)(nx + p) = 0 \qquad (l \neq 0, n \neq 0)$$

as $x = -m/l$ and $x = -p/n$.

Outcomes

After studying this section, you should be able to:

- ◇ solve a quadratic equation (or state there is no solution) by means of graphical and tabular methods (Activities 20, 21);
- ◇ solve a quadratic equation (or state there is no solution) by means of algebraic methods (Activities 22, 23, 24, 25, 26);
- ◇ explain your methods to others (Activity 27).

4 To catch a falling car

Aims The aims of the section are to consolidate the idea of a position–time graph model for motion in one dimension and the concepts of velocity and acceleration. It aims to demonstrate that motion with constant acceleration, like that of falling objects, produces a position–time graph which is parabolic (that is, that position is a quadratic function of time). It also aims to develop your skill in solving modelling problems involving quadratic functions and in solving problems about motion with constant acceleration. ◇

4.1 Falling objects

Objects dropped from a height above the ground fall under the influence of gravity; and, as long as air resistance is ignored, all bodies fall with the same constant acceleration, namely the acceleration due to gravity. The video band 'To catch a falling car', which you will be asked to watch shortly, repeats Galileo's experiment to show this, but using cars driven off the top of a skyscraper rather than cannon balls dropped from the leaning tower of Pisa.

When modelling the motion of a falling object, where the height from which it is dropped may vary, it is common to use a parameter for the various possible drop heights. So a general model for falling objects has the object falling from a given height H.

The numerical value of the acceleration due to gravity varies very slightly in different parts of the world, but for any one place it is constant. So, in general models for falling objects, it too is commonly represented by a parameter, namely g. The parameter g has a numerical value of about 9.81 metres per second per second (ms^{-2}), which is sometimes approximated to $10\,\text{ms}^{-2}$. As conventionally the y-axis points upwards, this is taken as the positive direction. However, since the object falls downwards and gravity acts downwards, both velocity and acceleration are negative. Therefore in such models the acceleration is represented by $-g$.

The velocity at the point of release is 0 (the object simply falls rather than being thrown). Hence, the velocity–time graph looks like Figure 22(a). The equation of this line is

$$v = -gt \qquad (t \geq 0, v \geq 0) \tag{9}$$

where v represents velocity and t represents time. The gradient $-g$ of this line is the acceleration. The position–time graph, which looks like Figure 22(b), is part of a parabola whose equation is:

The derivation of this equation is beyond the scope of this course.

$$h = H - \tfrac{1}{2}gt^2 \qquad (t \geq 0, H \geq h \geq 0) \tag{10}$$

The variable h gives the height above the ground at any time t from when the object is dropped from height H until it hits the ground.

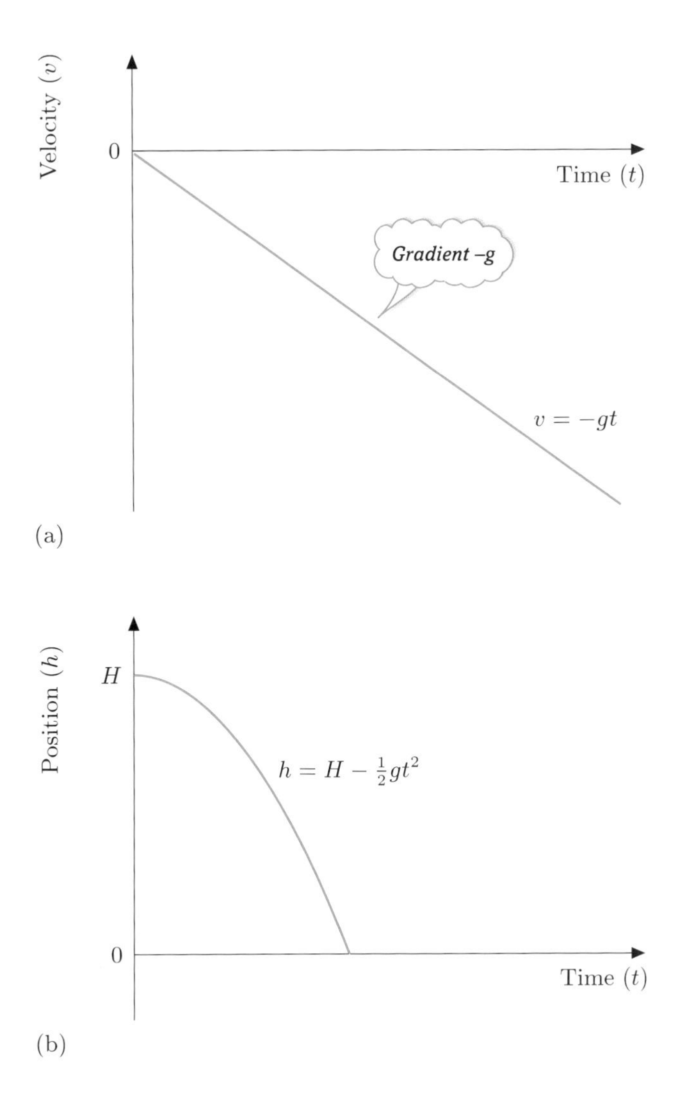

Figure 22

Equation (10) can be used not only to find the height at any given time but also to find at what time t the falling object is at height h. This involves some algebraic manipulation to make t the subject of the equation. Equation (10) is:

$$h = H - \tfrac{1}{2}gt^2$$

First, add $\frac{1}{2}gt^2$ to both sides and subtract h from both sides to get the term involving t^2 on its own on the left-hand side, giving:

$$\tfrac{1}{2}gt^2 = H - h$$

Now get the t^2 on its own by dividing both sides by $\frac{1}{2}g$, giving:

$$t^2 = \frac{2(H - h)}{g}$$

Now take square roots to get t. Since $t \geq 0$, the negative part of the parabola is not appropriate, and so neither is the negative square root solution. Hence:

$$t = \sqrt{\frac{2(H-h)}{g}} \qquad (11)$$

This is a general solution for the time t taken to reach a height h, in terms of a parameter H representing the height from which the object is dropped and g the acceleration due to gravity. In a particular situation, specific values would be substituted in for H and g.

In many situations, g can be taken as $10\,\mathrm{ms}^{-2}$. So with H and h in metres and t in seconds, the general function in equation (10) becomes $h = H - 5t^2$. The gradient of this function, based on equation (9), is given by the function $v = -10t$. Equation (11) then gives the time to reach a height h metres, in seconds, as $t = \sqrt{(H-h)/5}$.

Example 5 How deep is your well?

How long would it take for a stone to drop down a 10 m well?

If you measure distance from the bottom of the well, you can take H as 10 m. You want to find the time t when $h = 0\,\mathrm{m}$. So, taking g as $10\,\mathrm{ms}^{-2}$ equation (11) becomes:

$$t = \sqrt{\frac{2(10-0)}{10}} = \sqrt{2} = 1.4 \text{ seconds} \quad \text{(correct to one decimal place)}$$

This model ignores air resistance, so it would probably take a bit longer—of the order of one and a half seconds perhaps.

Time is often used colloquially instead of distance, for example 'How far is it?' 'It's a thirty-five minute drive.'

Activity 28 How many seconds for a 20 m well?

Use equation (11) with $g = 10\,\mathrm{ms}^{-2}$ to estimate how long it would take for a stone to drop down a 20 m well. Explain why this is *not* simply twice as long as for a 10 m well.

Activity 29 How long can you free fall?

(a) Suppose some parachutists are planning to let themselves fall out of a plane at a height of 1000 m and free fall to a height of 500 m before pulling the cords to open their parachutes. Use equation (11) with $g = 10\mathrm{ms}^{-2}$ to predict how many seconds in free fall this will be.

(b) The model given by equation (11) ignores air resistance. How would this affect the accuracy of the estimate you obtained in part (a)? Are there any other important assumptions which should be taken into consideration in interpreting the solution?

Now watch band 8 of DVD00107.

4.2 Video summary

The video showed the Volvo advertisement of a car driving off a skyscraper and asked the question: 'Taking this at its face value, as a record of something that happened, how did they work out when to start inflating the airbag?' To answer the question, you need to know how long it takes for the car to fall to the point where it will hit the inflated airbag, and how long it takes the airbag to inflate fully.

The video demonstrated that the position–time graph for a freely falling object is parabolic, implying that the acceleration is constant (equal to g).

The Volvo advertisement problem was to find t if $h = H - \frac{1}{2}gt^2$, where H is the height of the building and h is the height of the airbag when fully inflated. If the time taken to inflate the airbag is known, then the time at which inflation must start after the car starts its descent is obtained by subtracting the inflation time from the solution to the equation. For the given values of h, H and g you can solve the equation either by algebra or by using your calculator. This is not done on the video: but the results are given for particular numerical values.

The design of how a car safety airbag should open is related to the Volvo advertisement problem. The car safety airbag problem is one of calculating how quickly an airbag must inflate if it is to be of any use in an accident. For the driver's airbag, this involves modelling the motion of the driver's head relative to the steering wheel. If a constant-acceleration model is assumed, then this is equivalent to the falling car problem turned on its side, with the relevant parameters changed. In effect, it involves regarding the car body as stationary and the occupants of the car as accelerating relative to it.

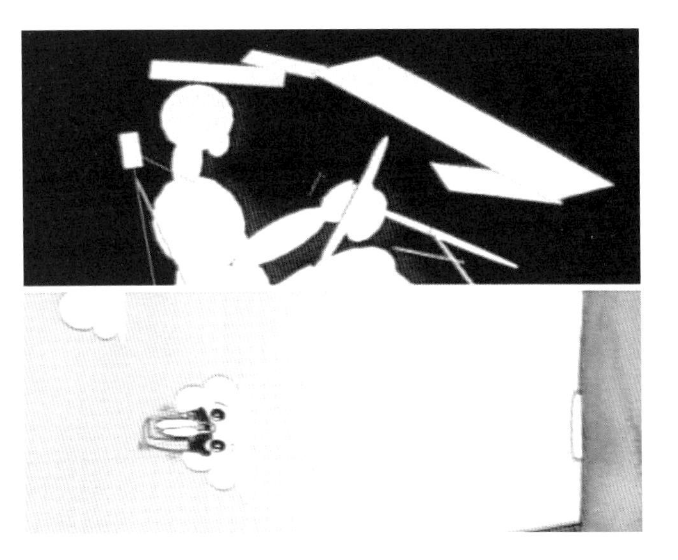

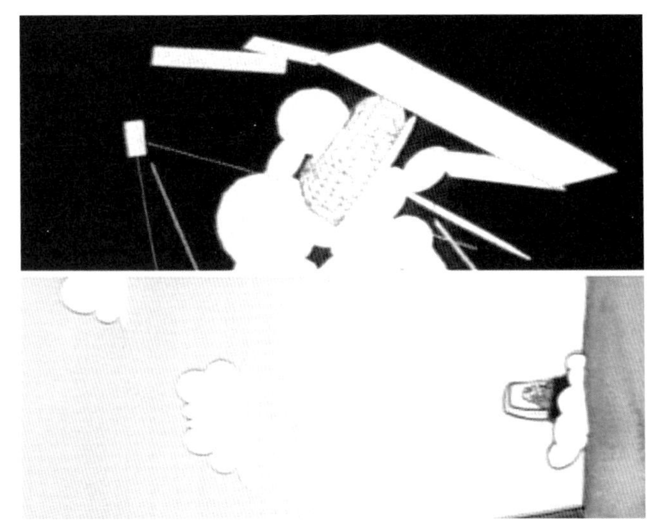

The video explained the major features of the operation of the airbag, including the fact that it begins to deflate as soon as it has reached its maximum size, so that the windscreen is obscured for as little time as possible. So it is crucial to get the timing right. The data required to solve the problem are the relevant dimensions of the car, the acceleration of the driver's head relative to the steering wheel, the acceleration of the passenger's head relative to the dashboard, and the size of the airbag when fully inflated.

4.3 Post-video work

Activity 30 Explaining airbags

Write down an explanation, as if to a friend, of how a safety airbag protects the driver's head and why it is crucial that it opens at exactly the right instant.

The data for the driver's head used in the video were: typical acceleration is $20g$ (twenty times greater than that due to gravity), the distance from the driver's head to the steering wheel is 550 mm and the desired final position of the driver's head is 250 mm from the steering wheel. The question addressed is: if the airbag is designed to fire 30 milliseconds after impact, will this protect the driver's head sufficiently in the event of a crash? The variables h and H are now distances from the steering wheel rather than heights above the ground. The calculation of the timing for the motion of the driver's head and the calculation of the timing for the airbag to be fully inflated, from the video, are repeated below.

The acceleration is $-20g$ instead of $-g$. So the equation of motion is:

$$h = H - 10gt^2 \tag{12}$$

The initial position of the driver's head is $H = 550\,\text{mm} = 0.55\,\text{m}$. The desired final position of driver's head is $h = 250\,\text{mm} = 0.25\,\text{m}$.

Substituting for h, H and g $(= 9.81\,\text{ms}^{-2})$ in equation (12) gives:

$$0.25 = 0.55 - 10 \times 9.81 \times t^2 = 0.55 - 98.1t^2$$

Subtracting 0.25 from both sides of the equation gives:

$$0 = 0.3 - 98.1t^2$$

Adding $98.1t^2$ to both sides gives:

$$98.1t^2 = 0.3$$

Dividing both sides by 98.1 gives:

$$t^2 = \frac{0.3}{98.1}$$

Taking the square root gives:

$$t = \pm\sqrt{\frac{0.3}{98.1}} = \pm 0.0553 \text{ seconds} \qquad \text{(to three significant figures)}$$

$$= \pm 55.3 \text{ milliseconds} \qquad \text{(to three significant figures)}$$

Only the positive solution is relevant here. So the airbag must be fully inflated by 55.3 milliseconds after the crash.

The airbag fires after 30 milliseconds, and so has about 25 milliseconds in which to become fully inflated.

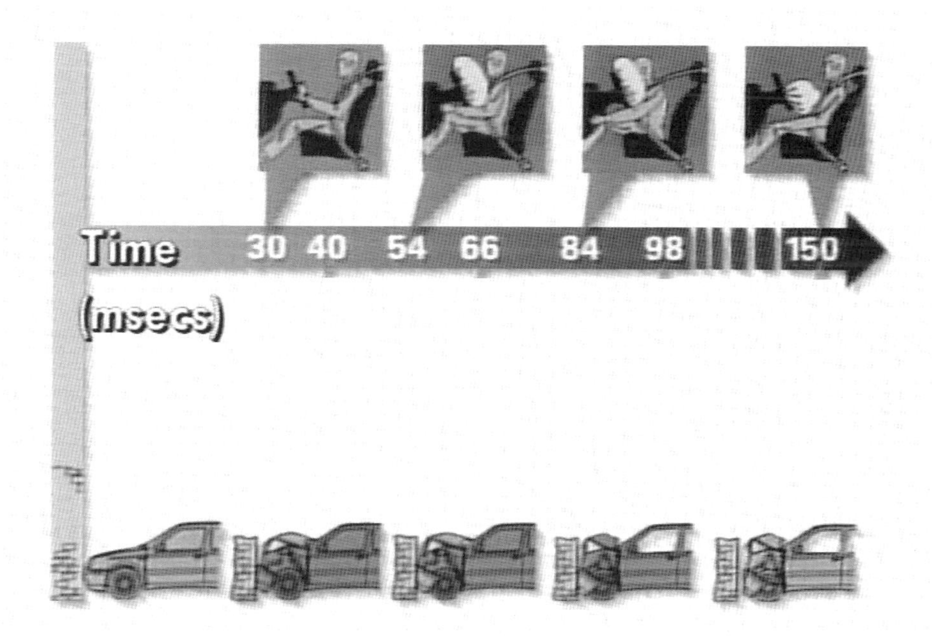

You may find it useful to view the last section of the video again before tackling the next activity.

Activity 31 The passenger's head

The data for the *passenger's* head is slightly different from that for the driver's head. The passenger's head is 750 mm from the dashboard of the car initially and its desired final position is 250 mm from the dashboard. By performing calculations similar to those above, predict when the passenger's airbag should be fully inflated.

In summary, quadratic functions are useful for modelling falling objects and other motion where acceleration can be modelled as constant, for example the motion of a driver's head relative to a car after a crash. However, such models ignore some aspects (for example, air resistance) and stress others (for example, the average acceleration being constant), and the results need to be interpreted in the light of this. Such models can be used in the design of safety features in vehicles, such as airbags.

Outcomes

After studying this section, you should be able to:

◇ use a constant acceleration model for the motion of falling objects and the heads of vehicle occupants relative to the vehicle in a crash (Activities 28–31).

Unit summary and outcomes

When describing motion, it is important to choose which direction is positive. Position, velocity and acceleration can then be positive or negative. Velocity is the rate of change of position, and acceleration is the rate of change of velocity.

This unit looked at some situations modelled by constant acceleration, including the acceleration due to gravity and the motion of car occupants' heads relative to the steering wheel or dashboard after a crash. In such situations, a constant-acceleration model leads to a straight-line model for velocity and a quadratic model for position.

Quadratic functions are useful in other situations too, for example modelling how population density varies with distance from a town centre. They are generally useful for quantities which have one maximum or one minimum point.

Quadratic models can be fitted to data points using the quadratic regression facility on your calculator. The calculator can also be used to find the gradient function of a quadratic function, confirming that it is a straight line.

The graph of a quadratic function is a parabola. It has an axis of symmetry and a vertex where the parabola meets this axis. If the axis of symmetry is the y-axis and the vertex is at the origin, then the equation of the parabolic graph is $y = ax^2$. When a is positive, the parabola opens upwards and the vertex gives a minimum value; when a is negative, it opens downwards, and the vertex is a maximum point. If the vertex is moved to the point (k, l) the equation becomes $y = a(x - k)^2 + l$.

In order to find when a quadratic function takes a particular value, you need to solve a quadratic equation. There are a number of ways to do this. You can use the facilities of your calculator or the formula, factorization or square root methods as appropriate. Sometimes there is no solution (the function never takes the particular value); there is one solution if it takes the particular value at the vertex; there are two solutions otherwise.

Activity 32 Reviewing quadratic functions

Think about what you know about quadratic functions. For each of the following, list those activities in this unit which demonstrate that you understand:

(a) the relevant technical terms;

(b) methods for solving quadratic equations;

(c) how to use quadratic functions in solving a problem.

Outcomes

Now you have finished this unit, you should be able to:

◇ use, and explain the meaning of, the following terms: 'position', 'velocity', 'acceleration', 'position–time graph', 'velocity–time graph', 'tangent to a curve', 'gradient of a curve', 'parabola', 'vertex' (of a parabola), 'axis' (of a parabola), 'quadratic function', 'parameters' (in the general form of a quadratic function), 'quadratic equation';

◇ sketch a velocity–time graph corresponding to a given position–time graph, and describe the acceleration involved;

◇ write down the formula for a parabola with axis parallel to the y-axis and with a given vertex;

◇ fit a parabola to (a) two or more data points, one of which is the vertex, or (b) a set of three or more data points, using quadratic regression on your calculator;

◇ find the gradient of the graph of a function, in particular of a parabola;

◇ solve quadratic equations or state that no real-number solutions exist;

◇ use quadratic functions for modelling motion and in other contexts.

Appendix

The material in all three sections of this Appendix is *optional*.

1 Two forms of the general equation of a parabola

A parabola with its vertex at (k, l) has the form:

$$\begin{aligned}(y-l) &= a(x-k)^2 = a(x-k)(x-k)\\ &= a\left(x(x-k) - k(x-k)\right) = a(x^2 - kx - kx + k^2)\\ &= a(x^2 - 2kx + k^2) = ax^2 - 2akx + ak^2\end{aligned}$$

Therefore:

$$y = ax^2 - 2akx + ak^2 + l$$

or equivalently:

$$y = ax^2 + bx + c \quad \text{where } b = -2ak \text{ and } c = ak^2 + l$$

The sign of a determines which way up the parabola is.

The vertex is at (k, l).

Since $b = -2ak$, then $k = \dfrac{b}{-2a} = -\dfrac{b}{2a}$.

Since $c = ak^2 + l$, then $l = c - ak^2$, and substituting $k = -\dfrac{b}{2a}$ into this equation gives:

$$l = c - \frac{ab^2}{4a^2} = c - \frac{b^2}{4a}$$

So the vertex is at $\left(-\dfrac{b}{2a}, c - \dfrac{b^2}{4a}\right)$.

2 Factorization

Consider, first, quadratic equations where the coefficient of x^2 is 1; that is, equations of the form

$$x^2 + bx + c = 0$$

where b and c are constants (not necessarily positive). If the equation can be factorized, in the form $(x+m)(x+n) = 0$, then the multiplied-out form of $(x+m)(x+n)$, namely:

$$x^2 + mx + nx + mn = x^2 + (m+n)x + mn$$

must be exactly the same as:

$$x^2 + bx + c$$

So $m + n$ must equal b, and mn must equal c.

So, to factorize $x^2 + bx + c$, you need to use a guess-and-try method to find two numbers m and n which (taking note of their signs) *multiply* to c and *add* to b. The *factors* are then $(x + m)$ and $(x + n)$.

Example 6

Use factorization to solve the following quadratic equation:

$$x^2 - 5x + 6 = 0$$

The aim is to find two numbers m and n so that:

$$(x + m)(x + n) = x^2 - 5x + 6$$

This means that $mn = 6$ and $m + n = -5$.

Look first at the possibilities which give $mn = 6$. In terms of whole numbers, the possibilities are: 1 and 6; -1 and -6; 2 and 3; or -2 and -3.

Testing each combination to see if they add to -5 shows that the only possibility is -2 and -3.

Here the numbers -3 and -2 *multiply* to give $+6$ and *add* to give -5. So the factors are $(x + (-3))$ and $(x + (-2))$, that is $(x - 3)$ and $(x - 2)$. So the given quadratic equation factorizes to

$$(x - 3)(x - 2) = 0$$

and has the following solutions:

$$x = 3 \quad \text{and} \quad x = 2$$

Check that these are indeed solutions:

$$\text{if } x = 2, \text{ then } x^2 - 5x + 6 = 2^2 - 5 \times 2 + 6 = 4 - 10 + 6 = 0$$
$$\text{if } x = 3, \text{ then } x^2 - 5x + 6 = 3^2 - 5 \times 3 + 6 = 9 - 15 + 6 = 0$$

Example 7

Use factorization to solve the following quadratic equation:

$$x^2 + 6x + 9 = 0$$

The possible pairs of whole numbers which multiply to give 9 are: 1, 9; -1, -9; 3, 3; -3, -3.

Only the pair of numbers $+3$ and $+3$ *multiply* to give $+9$ and at the same time *add* to give $+6$. So the factors are $(x + 3)$ and $(x + 3)$. Here the given quadratic equation factorizes to

$$(x + 3)(x + 3) = 0$$

and the only solution is

$$x = -3$$

because the two identical brackets lead to just one solution.

Check that this is indeed a solution:

$$\text{if } x = -3, \text{ then } x^2 + 6x + 9 = (-3)^2 + 6 \times (-3) + 9 = 9 - 18 + 9 = 0$$

The numbers m and n in the factors $(x + m)$ and $(x + n)$ need not be whole numbers. In particular, this often happens when the coefficient of x^2 is not 1. In such a case, at least one of the two numbers may be a fraction.

When the coefficient of x^2 is not 1, the first step in the factorization process can be to divide through by the coefficient of x^2.

Example 8

Use factorization to solve the following quadratic equation:

$$2y^2 - 9y - 5 = 0$$

First divide through by 2, to obtain:

$$y^2 - \tfrac{9}{2}y - \tfrac{5}{2} = 0$$

Then look for a pair of numbers which *multiply* to give $-\frac{5}{2}$ and *add* to give $-\frac{9}{2}$. The numbers are -5 and $\frac{1}{2}$. (Whenever the constant term is negative, one number will be positive and the other negative.) The factors are therefore $(y - 5)$ and $(y + \frac{1}{2})$ and the equation factorizes to the following:

$$(y - 5)(y + \tfrac{1}{2}) = 0$$

This gives the following solutions:

$$y = 5 \quad \text{and} \quad y = -\tfrac{1}{2}$$

Check that these are indeed solutions:

$$\text{if } y = 5, \text{ then } 2y^2 - 9y - 5 = 2 \times 5^2 - 9 \times 5 - 5 = 50 - 45 - 5 = 0$$
$$\text{if } y = -\tfrac{1}{2}, \text{ then } 2y^2 - 9y - 5 = 2 \times (-\tfrac{1}{2})^2 - 9 \times (-\tfrac{1}{2}) - 5 = \tfrac{1}{2} + \tfrac{9}{2} - 5 = 0$$

Note that by no means all quadratic equations will factorize using whole numbers or fractions. If, therefore, the factors are not quickly apparent, use the formula method.

3 Derivation of the formula for solving quadratic equations

From the text you know that

$$(x + m)^2 = n \quad (n \geq 0) \qquad (13)$$

has the following solutions:

$$x = -m \pm \sqrt{n}$$

Multiplying out equation (13) gives:

$$x^2 + 2mx + m^2 = n$$

or equivalently:

$$x^2 + 2mx + (m^2 - n) = 0 \qquad (14)$$

The quadratic equation

$$ax^2 + bx + c = 0 \qquad (a \neq 0) \qquad (15)$$

can first be divided throughout by a and then compared with equation (14). As the solution of equation (14) is known, the solution of equation (15) can be found by comparing the terms.

Dividing equation (15) by a gives:

$$x^2 + \frac{b}{a}x + \frac{c}{a} = 0$$

Comparison with equation (14) gives:

$$2m = \frac{b}{a} \qquad \text{(from the coefficients of } x\text{);}$$

$$m^2 - n = \frac{c}{a} \qquad \text{(from the constant terms).}$$

Hence $m = \dfrac{b}{2a}$. Also, since $n = m^2 - \dfrac{c}{a}$, substituting $m = \dfrac{b}{2a}$ gives:

$$n = \left(\frac{b}{2a}\right)^2 - \frac{c}{a} = \frac{b^2}{4a^2} - \frac{c}{a} = \frac{b^2 - 4ac}{4a^2}$$

So the solution to equation (15) (which is $x = -m \pm \sqrt{n}$) is

$$\begin{aligned} x &= -\frac{b}{2a} \pm \sqrt{\left(\frac{b^2 - 4ac}{4a^2}\right)} \\ &= -\frac{b}{2a} \pm \frac{\sqrt{(b^2 - 4ac)}}{2a} \\ &= \frac{-b \pm \sqrt{(b^2 - 4ac)}}{2a} \end{aligned}$$

which is the claimed formula.

Comments on Activities

Activity 1

(a) Obviously your description will be personal to you. But you should have included the following points.

(i) The outward journey: a distance of 10 km, your position has changed—you are 10 km from home.

(ii) The return journey: also a distance of 10 km, your position has changed—you are back where you started.

(b) Possible distance–time and position–time graphs are shown in Figures 23 and 24.

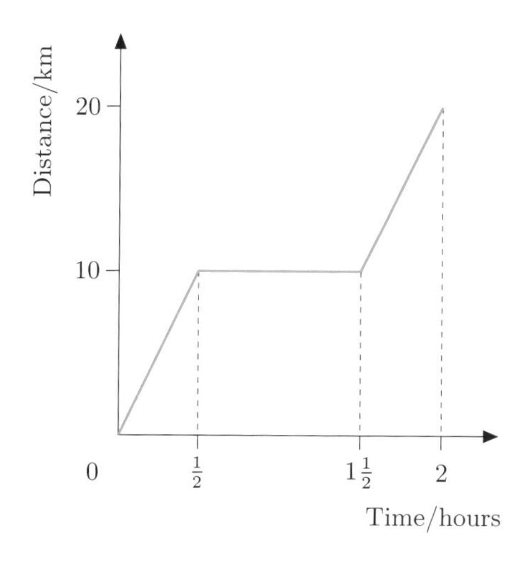

Figure 23

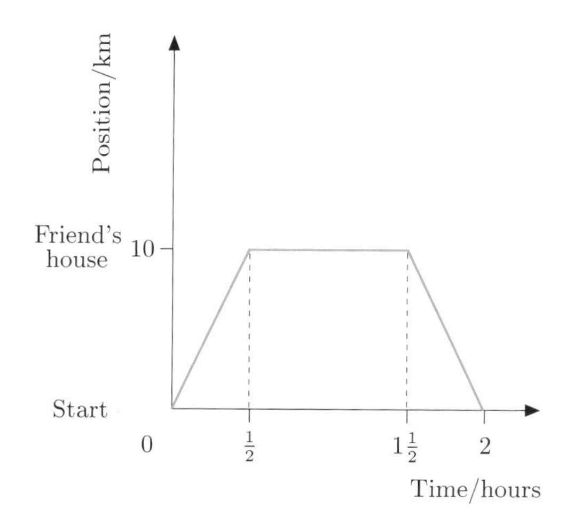

Figure 24

Your graphs may be different, depending upon how much detail you included about the journey, for example whether you considered speeding up and slowing down, and upon how you measured position. Figure 24 measures position from the starting point. If you had measured position from your friend's house, then your graph of position would be different, something like that in Figure 25.

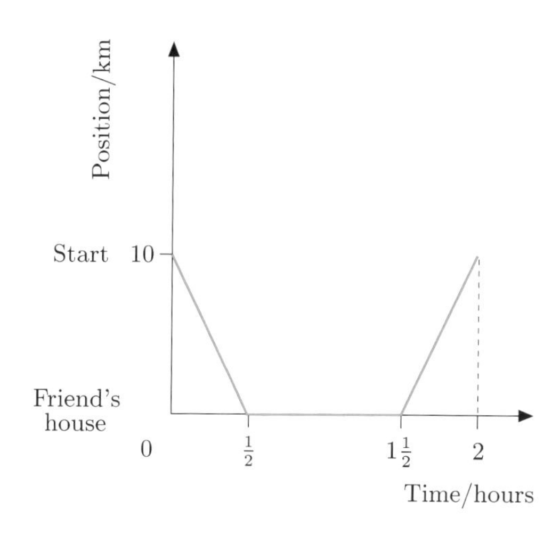

Figure 25

Your descriptions embody models of the trip. You have probably stressed the distances and positions along the bus route and directions towards or away from your starting point or your friend's house. You have probably ignored aspects of the journey such as bends in the road, corners, getting from the bus stop to your friend's house, and all the times the bus had to speed up or slow down due to traffic lights, other traffic, setting down or picking up passengers, and so on.

Activity 2

See the comments in the main text after the activity.

Activity 3

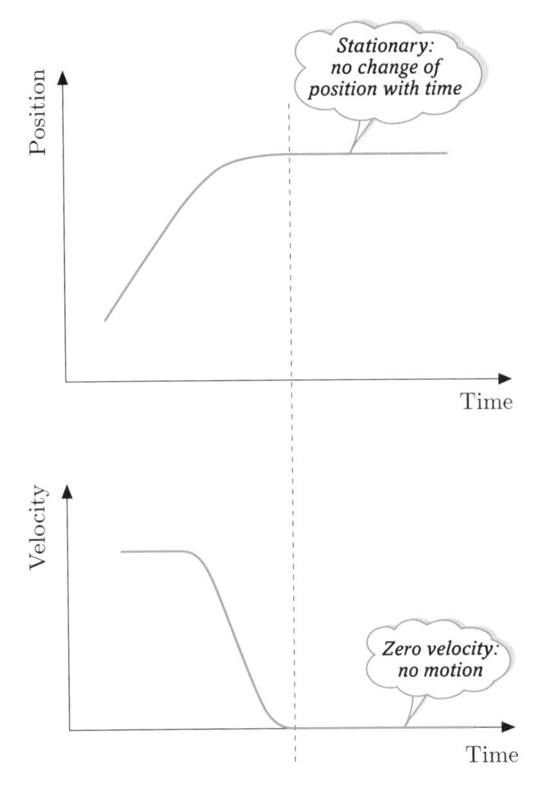

Figure 26

Activity 4

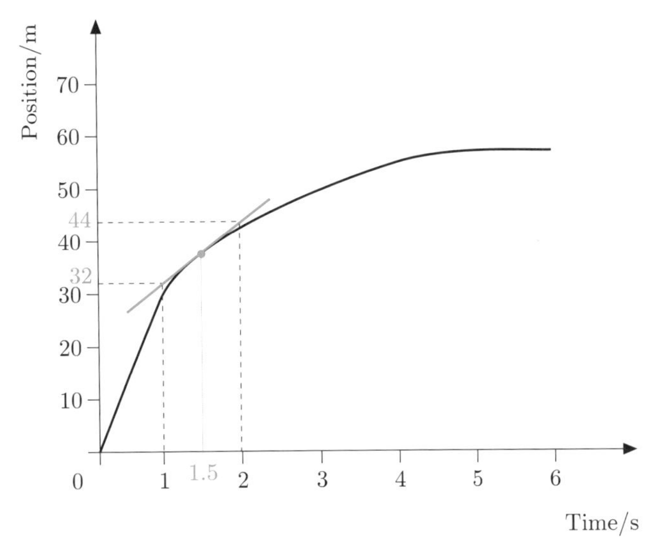

Figure 27

Your answer may differ slightly from that given, if you have drawn the tangent differently.

For the tangent shown in Figure 27, going through the points $(1, 32)$ and $(2, 44)$, the gradient is:

$$\frac{44\text{ m} - 32\text{ m}}{2\text{ s} - 1\text{ s}} = 12\,\text{ms}^{-1}$$

Hence the velocity is 12 metres per second after 1.5 seconds.

(Note that it is helpful, when computing gradients, to choose points on the tangent that give a convenient denominator, such as 1 s in this case.)

Activity 5

The moving bus has positive velocity. As it approaches the bus stop, it slows down; so its velocity (v) decreases, which means that the acceleration (a) is negative. When the bus is stationary, both velocity and acceleration are zero. As the bus starts off, its velocity increases in the forward direction, and so the acceleration is positive, as is the velocity.

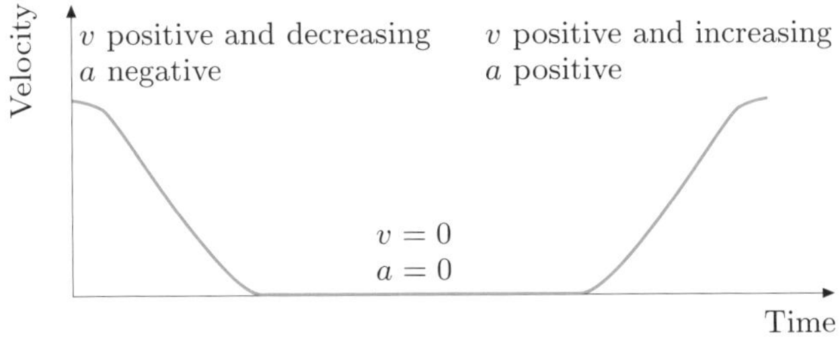

Figure 28

Activity 6

In (a), (b) and (d), the velocity is changing and so there is acceleration. In (c) and (e), the velocity is constant, and so the acceleration is zero.

Activity 7

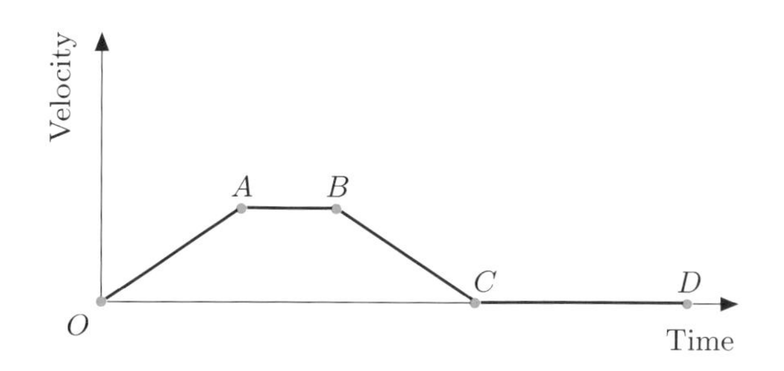

Figure 29

OA: The slope of the graph (giving the velocity) is getting steeper. So the velocity increases (from zero) in the positive direction, and the acceleration is positive.

AB: The slope of the graph is constant. So the velocity is constant, and the acceleration is zero.

BC: The slope of the graph is getting less steep. So the velocity decreases (to zero), and the acceleration is negative.

CD: The slope of the graph is zero. So the velocity is zero, and the acceleration is zero.

Activity 8

Obviously your answer will be in your own words, but it should cover the following.

The gradient or rate of change of a position–time graph gives the velocity. The gradient or rate of change of a velocity–time graph gives the acceleration. Increasing velocity in the positive direction means a positive acceleration and decreasing velocity in the positive direction means a negative acceleration.

Your answer might have made use of diagrams like those for Activity 7.

Activity 9

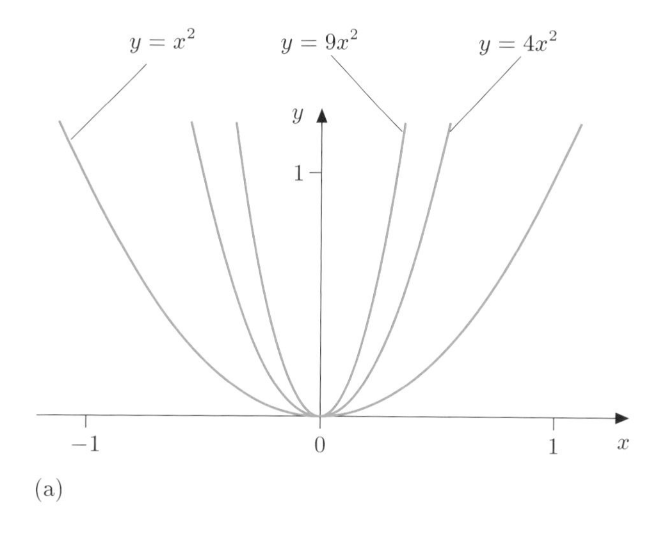

(a)

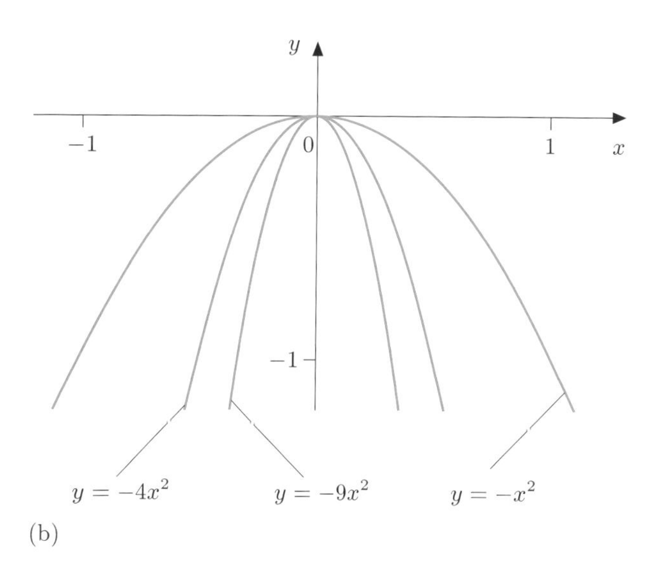

(b)

Figure 30

If a is positive, the parabola opens upwards; if a is negative, it opens downwards. The size of a determines how widely the parabola opens: the smaller is a, the more widely it opens.

Activity 10

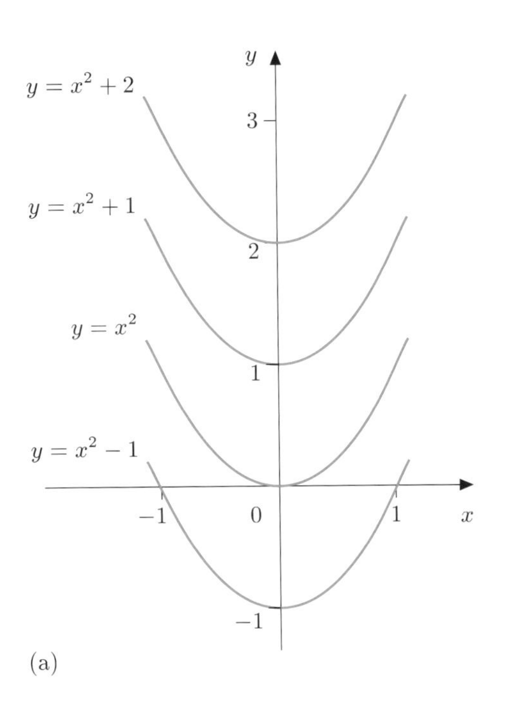

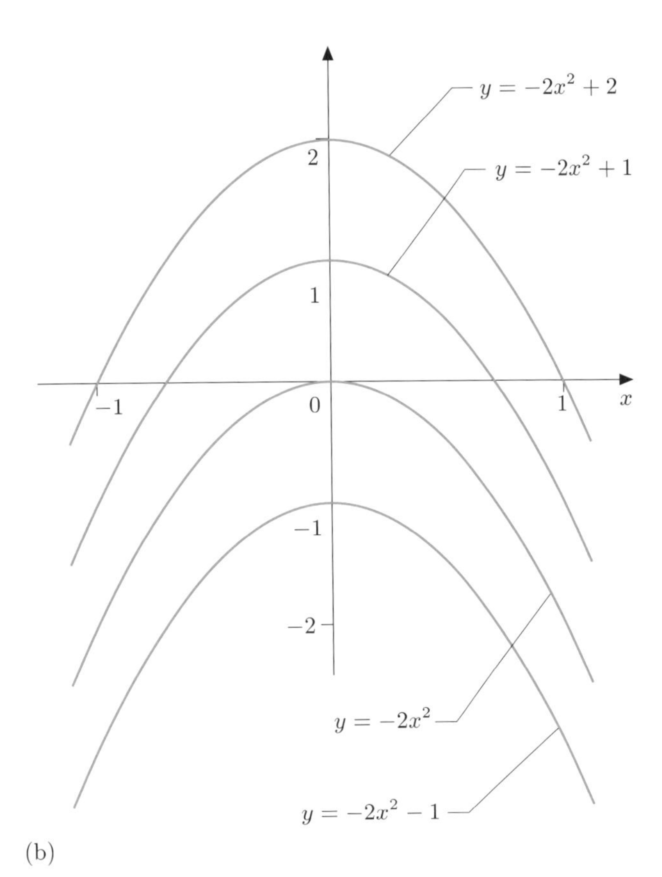

Figure 31

The position of the vertex changes with l. The value of l gives its y-coordinate: the vertex is at $(0, l)$.

Activity 11

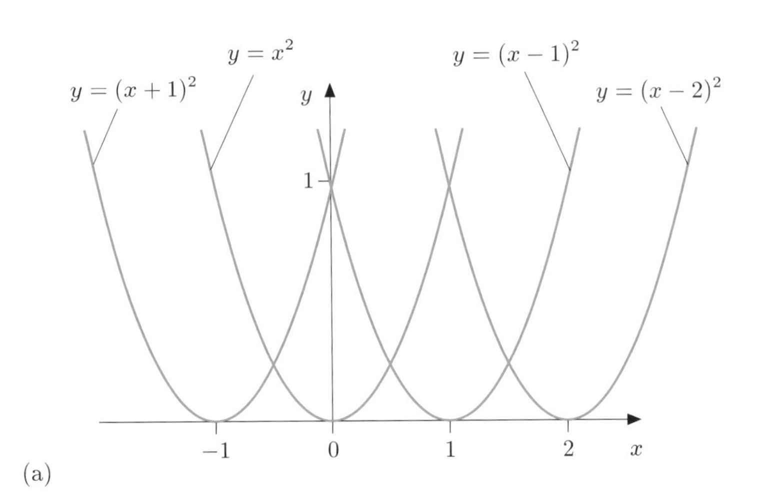

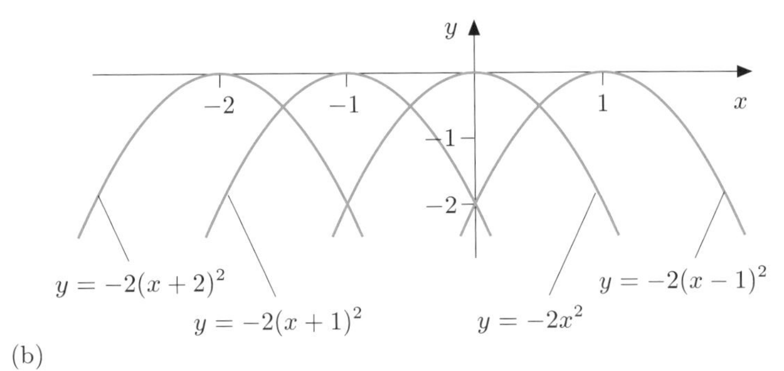

Figure 32

The value of k determines how far to the left or right of the y-axis the parabola is. The vertex of $y = a(x-k)^2$ is at $(k, 0)$.

Activity 12

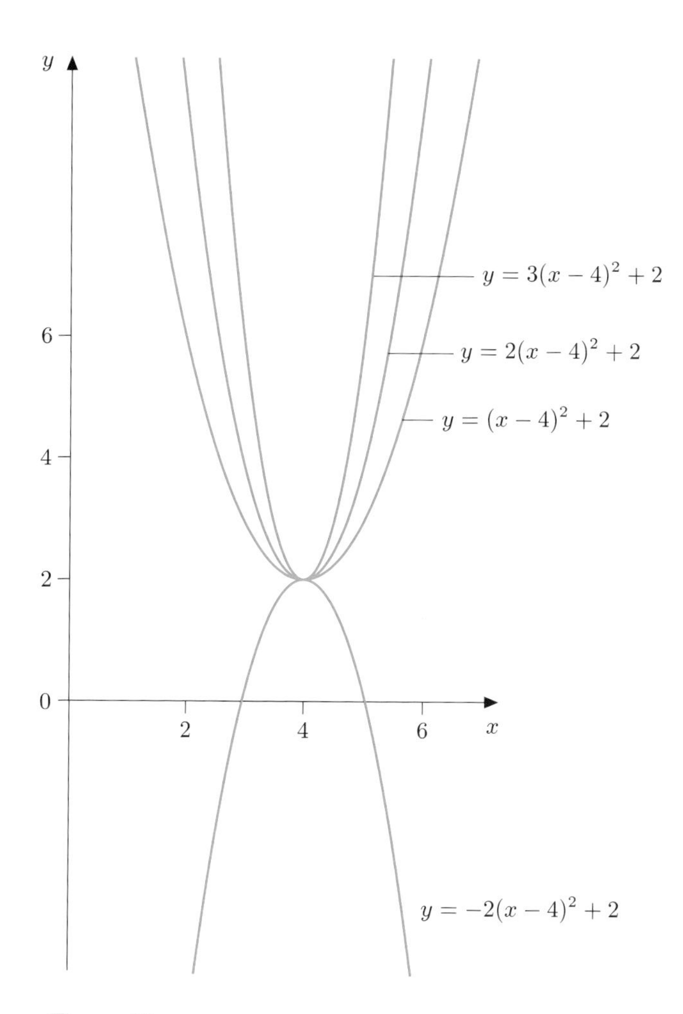

Figure 33

Activity 13

(a) (i) $y = a(x-3)^2 + 0$

that is:

$$y = a(x-3)^2$$

(ii) $y = a(x-(-2))^2 + (-1)$

that is:

$$y = a(x+2)^2 - 1$$

(b) (i) $(9, -3)$

(ii) $(-6, 2)$

Activity 14

Using your calculator with quadratic regression gives the height (y metres) as a function of the time (x seconds) as

$$y = ax^2 + bx + c$$

where $a = -5$, $b = 5$ and $c = 0$. In other words:

$$y = -5x^2 + 5x$$

Plotting this gives the curve shown in Figure 34.

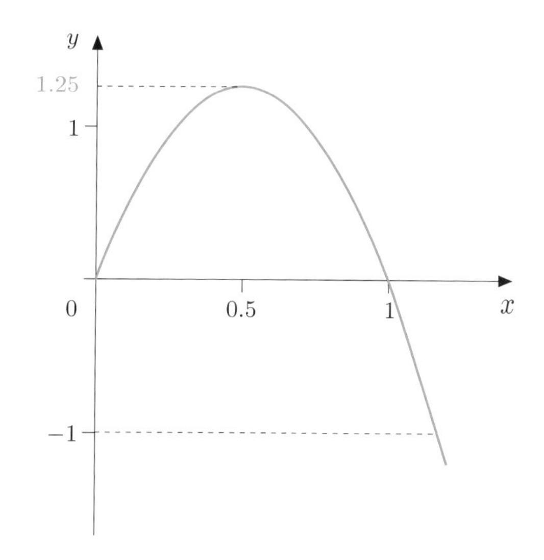

Figure 34

Using the trace and/or table facilities of the course calculator to find when y is -1 gives x as 1.17 (correct to two decimal places).

Activity 15

Your summary should have included most or all of the following points.

The axis of a parabola is a line of symmetry through the parabola. The point where the axis cuts the parabola is called the vertex.

A parabola with its axis along the y-axis and its vertex at the origin is produced by the following function:

$$y = ax^2 \qquad (a \neq 0)$$

If a is positive, the parabola opens upwards; if a is negative, the parabola opens downwards. The smaller the value of a, the more spread out the parabola.

If the parabola is shifted so the vertex is at (k, l) but the axis is still parallel to the y-axis, the equation is:

$$(y - l) = a(x - k)^2 \qquad (a \neq 0)$$

or equivalently:

$$y = a(x - k)^2 + l \qquad (a \neq 0)$$

An alternative form of this equation is:

$$y = ax^2 + bx + c \qquad (a \neq 0)$$

This specifies the general form of a quadratic function: the sign of a determines which way up the parabola is; the value of c gives the y-intercept of the parabola.

Parabolas with their axes parallel to the y-axis are particularly useful in modelling situations where some variable rises to a maximum value and then falls off, or falls to a minimum value and then rises again.

A parabola can be fitted to data points using a process called quadratic regression. In general, at least three data points are needed to fit a parabola.

Activity 16

(a) Quadratic regression gives $y = ax^2 + bx + c$ where $a = -157.684$, $b = 0.198$, $c = 0.994$ (all to three decimal places). So the function is as follows:

$$y = -157.684x^2 + 0.198x + 0.994$$

Call this model 1.

(b) If the vertex is $(0, 1)$, then the quadratic function is of the form:

$$y = ax^2 + 1$$

Since the function passes though the point $(0.08, 0)$, this gives:

$$0 = a(0.08)^2 + 1$$

Rearranging gives:

$$a = -\frac{1}{(0.08)^2} = -156.25$$

So the function is:

$$y = -156.25x^2 + 1$$

Call this model 2.

(c) The table facility on the calculator gives the following comparison (data points given to two decimal places, to match the original data).

Time	Distance (metres)		
(seconds)	data	model 1	model 2
0	1.00	0.99	1.00
0.01	0.97	0.98	0.98
0.02	0.93	0.93	0.94
0.03	0.87	0.86	0.86
0.04	0.75	0.75	0.75
0.05	0.61	0.61	0.61
0.06	0.43	0.44	0.44
0.07	0.24	0.24	0.23
0.08	0.00	0.00	0.00

Both functions fit the data points very well.

(d) Plotting both functions and the data points shows virtually no difference between the models.

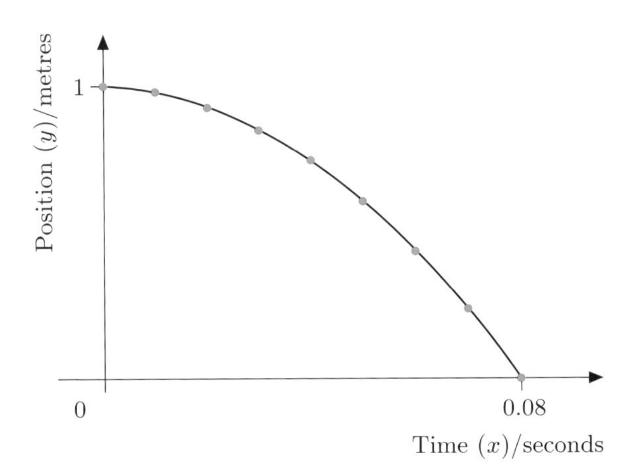

Figure 35

(e) The gradient at $x = 0.08$ (when the head hits the windscreen) according to both models is -25 (to two significant figures), which predicts a speed of $25\,\mathrm{m\,s^{-1}}$ as the head hits the windscreen. (The minus indicates that direction of motion is the negative direction.)

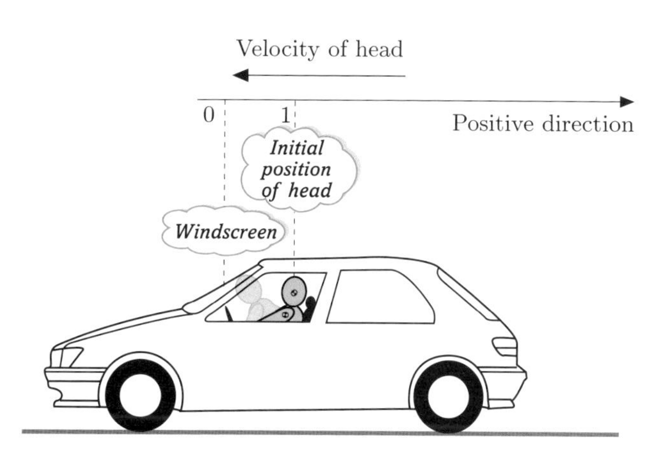

Figure 36

Activity 17

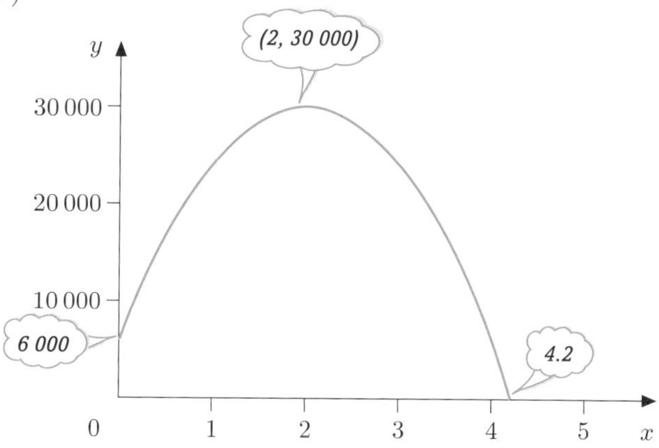

Distances are in km and population densities in number of people per km^2.

(a) The parabola goes through the points $(0, 6000)$, $(2, 30\,000)$ (vertex) and by symmetry $(4, 6000)$.

The equation from quadratic regression is:

$$y = -6000x^2 + 24\,000x + 6000$$

By algebra:

$$y = a(x-k)^2 + l$$

where $k = 2$ and $l = 30\,000$, so:

$$y = a(x-2)^2 + 30\,000$$

For $(0, 6000)$:

$$\begin{aligned} 6000 &= a(-2)^2 + 30\,000 \\ -24\,000 &= 4a \\ a &= -6000 \end{aligned}$$

Therefore:

$$y = -6000(x-2)^2 + 30\,000$$

which multiplies out to:

$$y = -6\,000(x^2 - 4x + 4) + 30\,000$$

and simplifies to:

$$y = -6000x^2 + 24\,000x + 6000$$

When $x = 0.5$, $y = 16\,500$, so the population density is predicted by this model to be 16 500 people per km^2 at 0.5 km from the centre.

(b)

Figure 37

(c) If the population density is y people per km^2 and the distance from town centre is x km, then the model is only valid for $y \geq 0$, $x \geq 0$.

Using the trace facility of the course calculator on the graph suggests that y becomes zero when x is about 4.2. So the model will be meaningless for distances greater than about 4 km from the centre.

Activity 18

The variables are sales per week and time (in weeks). Let y CDs per week be the sales in week x.

The first week is $x = 1$ and the sixth $x = 6$. So the given information is $y = 10\,000$ at $x = 1$ and $y = 22\,000$ at $x = 6$ (peak value).

A quadratic model is required. From the general equation of a parabola, y and x are related by

$$y = a(x-k)^2 + l$$

where (k, l) are the coordinates of the vertex.

Here $k = 6$ and $l = 22\,000$. So:

$$y = a(x-6)^2 + 22\,000$$

Also $y = 10\,000$ at $x = 1$. So:

$$\begin{aligned} 10\,000 &= a(1-6)^2 + 22\,000 \\ -12\,000 &= 25a \\ a &= -480 \end{aligned}$$

The equation is therefore:

$$y = -480(x - 6)^2 + 22\,000$$

This leads to the table below predicting future weekly sales.

Week (x)	Sales (y) (rounded to nearest hundred)
6	22 000 (as expected)
7	21 500
8	20 100
9	17 700
10	14 300
11	10 000
12	4 700
13	−1 500

Negative sales are not possible; so, according to this simple model, sales will fall to zero about 12 to 13 weeks after the CD first went on sale.

Activity 19

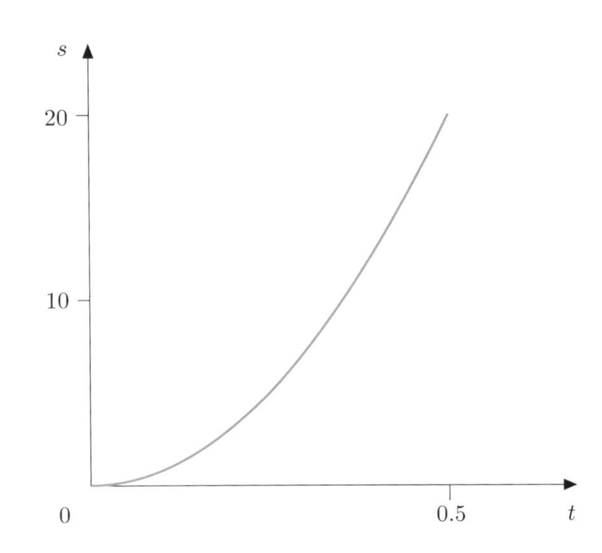

Figure 38 $s = 80t^2 \quad (0 \le t \le 0.5)$

t (mins)	s (km) (rounded to two decimal places)
0	0
10	2.22
20	8.89
30	20

The table difference was set to $\frac{1}{6}$ of an hour (10 minutes).

Activity 20

(a)

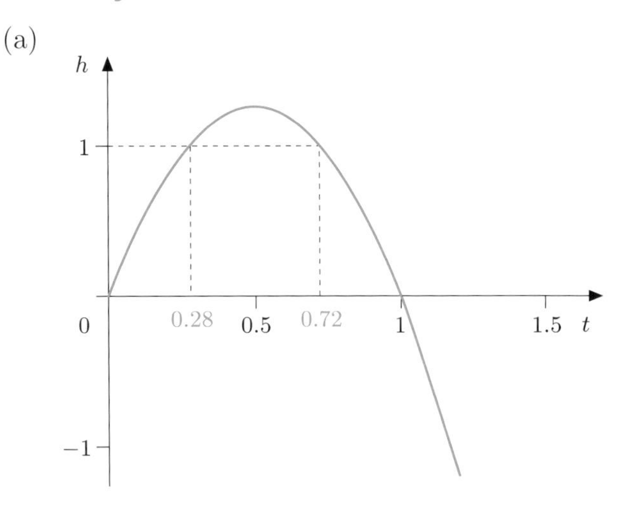

Figure 39

When $h = 1$, $t = 0.28$ and $t = 0.72$ (to two decimal places).

(b) For $t = 0.28$,
$y = 1.008 = 1.0$ (to two significant figures).
For $t = 0.72$,
$y = 1.008 = 1.0$ (to two significant figures).

(c) $h = 1.25$ at $t = 0.5$.
h never is 1.5 on this parabola.

Activity 21

You should find that the graph crosses the x-axis very close to $x = 0.17$ and $x = 3.83$.

Activity 22

(a) $x = \pm 5$
Check: $x = 5$ gives $x^2 = 25$;
$x = -5$ gives $x^2 = 25$.

(b) $x = \pm\frac{4}{3}$
Check: $x = +\frac{4}{3}$ gives $9x^2 = 9 \times \frac{16}{9} = 16$;
$x = -\frac{4}{3}$ gives $9x^2 = 9 \times \frac{16}{9} = 16$.

(c) $y = -1 \pm 4$, so $y = 3$ and $y = -5$.
Check: $y = 3$ gives $(3 + 1)^2 = 16$;
$y = -5$ gives $(-5 + 1)^2 = 16$.

(d) $x = 3 \pm \sqrt{15}$, so $x = 6.873$ and $x = -0.873$ (to three decimal places).
Check: $x = 6.873$ gives $(6.873 - 3)^2 = 15.00$;
$x = -0.873$ gives $(-0.873 - 3)^2 = 15.00$.

(e) This quadratic equation has no solutions because it gives $(z+2)^2 = -9$ and the square root of -9 is not a real number.

Activity 23

(a) This factorizes to

$$x(x+12) = 0$$

and has the following solutions:

$$x = 0 \quad \text{and} \quad x = -12$$

(b) This factorizes to

$$3y(y-3) = 0$$

and has the following solutions:

$$y = 0 \quad \text{and} \quad y = 3$$

(c) This factorizes to

$$z(5z+1) = 0$$

and has the following solutions:

$$z = 0 \quad \text{and} \quad z = -\tfrac{1}{5}$$

(d) One or other of the brackets must equal zero. So the solutions are:

$$x = 9 \quad \text{and} \quad x = -\tfrac{11}{2}$$

Activity 24

(a) Here $a = 1$, $b = -9$ and $c = 2$. The formula gives:

$$\begin{aligned} x &= \frac{-(-9) \pm \sqrt{[(-9)^2 - 4 \times 1 \times 2]}}{2 \times 1} \\ &= \frac{9 \pm \sqrt{73}}{2} \\ &\simeq \frac{9 \pm 8.544}{2} \end{aligned}$$

So the solutions are:

$$x = 8.772 \quad \text{and} \quad x = 0.228$$

(to three decimal places)

Check:
$(8.772)^2 - 9 \times 8.772 + 2$
$\simeq 76.948 - 78.948 + 2 = 0$
$(0.228)^2 - 9 \times 0.228 + 2$
$\simeq 0.052 - 2.052 - 2 = 0$

(b) Here $a = 2$, $b = 3$ and $c = -5$. The formula gives:

$$\begin{aligned} y &= \frac{-3 \pm \sqrt{[3^2 - 4 \times 2 \times (-5)]}}{2 \times 2} \\ &= \frac{-3 \pm \sqrt{(9 + 40)}}{4} \\ &= \frac{-3 \pm 7}{4} \end{aligned}$$

So the solutions are:

$$y = 1 \quad \text{and} \quad y = -2.5$$

Check:
$2 \times 1^2 + 3 \times 1 - 5 = 2 + 3 - 5 = 0$
$2 \times (-2.5)^2 + 3 \times (-2.5) - 5$
$= 12.5 - 7.5 - 5 = 0$

(c) Here $a = 4$, $b = 12$ and $c = 9$. The formula gives:

$$\begin{aligned} m &= \frac{-12 \pm \sqrt{(12^2 - 4 \times 4 \times 9)}}{2 \times 4} \\ &= \frac{-12 \pm \sqrt{0}}{8} \end{aligned}$$

So the only solution is $m = -1.5$.

Check:
$4 \times (-1.5)^2 + 12 \times (-1.5) + 9$
$= 9 - 18 + 9 = 0$

(d) Here $a = 5$, $b = -2$ and $c = 2$. The formula gives:

$$\begin{aligned} z &= \frac{-(-2) \pm \sqrt{[(-2)^2 - 4 \times 5 \times 2]}}{2 \times 5} \\ &= \frac{2 \pm \sqrt{(4 - 40)}}{10} \\ &= \frac{2 \pm \sqrt{-36}}{10} \end{aligned}$$

So this quadratic equation has no real-number solutions.

Check by drawing the graph on your calculator.

Activity 25

(a) The given quadratic equation can be divided through by 2000 to give:

$$-3x^2 + 12x - 2 = 0$$

Here $a = -3, b = 12$ and $c = -2$. The formula gives:

$$x = \frac{-12 \pm \sqrt{[12^2 - 4 \times (-3) \times (-2)]}}{2 \times (-3)}$$

$$= \frac{-12 \pm \sqrt{(144 - 24)}}{-6}$$

$$\simeq \frac{-12 \pm 10.954}{-6}$$

So the solutions are:

$$x = 3.83 \quad \text{and} \quad x = 0.17$$

(to two decimal places)

(You could have divided through by -2000. This would have left you fewer minus signs to deal with, but would, of course, have led to the same solutions.)

(b) $y \geq 10\,000$ whenever x lies between 0.17 and 3.83, as Figure 40 illustrates.

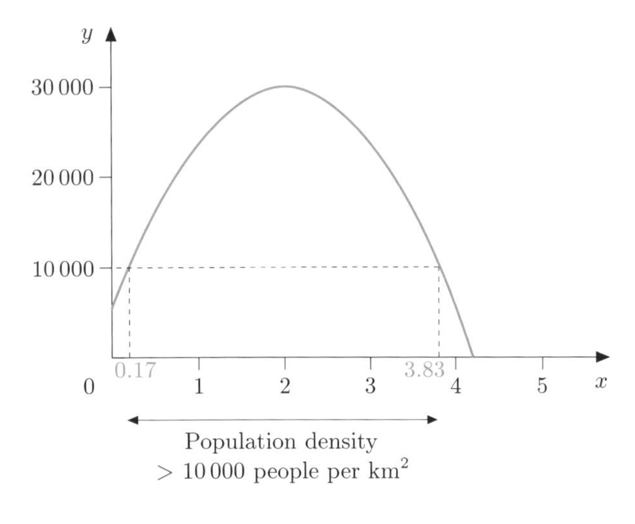

Figure 40

Therefore the model predicts that the population density has a value of at least 10 000 people per km^2 between 0.17 km and 3.83 km from the town centre.

(These are precise results correct to the nearest 10 metres, but since the equation arose out of some modelling assumptions, its results are only approximations and should be treated with caution.)

Activity 26

(a) $x = -2$ and $x = -3$.

(b) $y = 2$ and $y = 8$.

(c) $p = -0.268$ and $p = -3.732$ (to three decimal places).

(d) $x = 0.618$ and $x = -1.618$ (to three decimal places).

(e) This quadratic equation has no real-number solutions.

Activity 27

Check you have included all the important points when you read the summary that follows Activity 27 in the main text.

Activity 28

From equation (11):

$$t = \sqrt{\frac{40}{10}} = \sqrt{4} = 2$$

So it takes 2 seconds.

The stone is accelerating, so does not take twice as long—it travels faster over the second 10 m than the first 10 m.

Activity 29

(a) $H = 1000$ m and $h = 500$ m. So, taking $g = 10\,\text{ms}^{-2}$, equation (11) gives:

$$t = \sqrt{\frac{2(1000 - 500)}{10}} = \sqrt{100} = 10$$

So this model predicts 10 seconds of free fall.

(b) With air resistance, the fall will be slower, so 10 seconds is an underestimate.

The reaction times of both the parachutist and the parachute are also ignored.

Activity 30

Everyone's explanation will differ, but you should include the following points.

- ◇ The airbag must be fully inflated when the driver's head hits it, in order to cushion the impact against the steering wheel.
- ◇ The airbag deflates very fast after becoming fully inflated, in order for the windscreen to be obscured for the minimum time.
- ◇ The whole process takes a fraction of a second (less than a tenth of a second!).

Activity 31

The calculation is as for the driver but with H replaced by 750 mm = 0.75 m.

$$\begin{aligned} 0.25 &= 0.75 - 10 \times 9.81 \times t^2 \\ &= 0.75 - 98.1t^2 \\ 98.1t^2 &= 0.75 - 0.25 = 0.5 \\ t^2 &= \frac{0.5}{98.1} \\ t &= \sqrt{\frac{0.5}{98.1}} = 0.0714 \end{aligned}$$

(to three significant figures)

Hence the airbag must be open by about 70 milliseconds after the crash; that is, about 40 milliseconds after it fires.

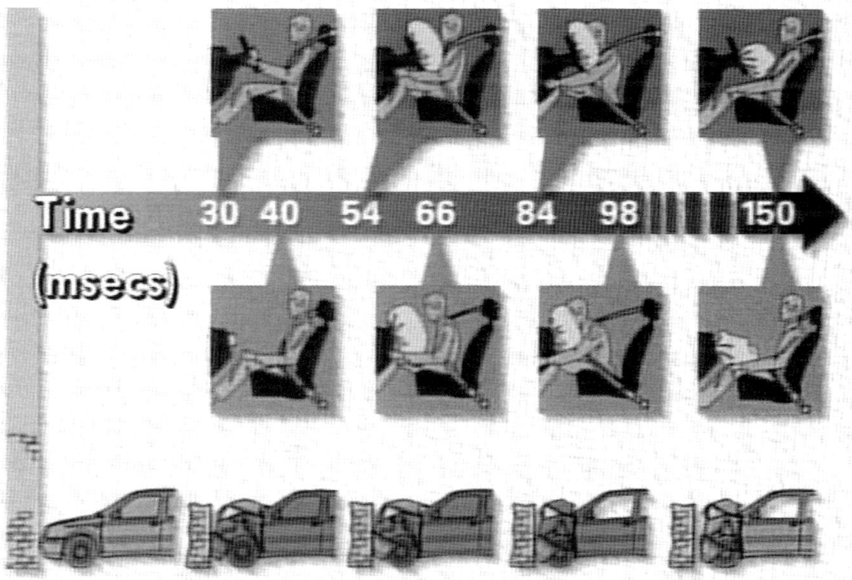

Activity 32

There is no comment on this activity.

Acknowledgements

Grateful acknowledgement is made to the following sources for permission to reproduce material in this unit.

Cover

John Regis, sprinter: Press Assocation; crowd scene: Camera Press; car prices: Edinburgh Mathematical Teaching Group; other photographs: Mike Levers, Photographic Department, The Open University.

Index